SpringerBriefs in Computer Science

Series Editors
Stan Zdonik
Peng Ning
Shashi Shekhar
Jonathan Katz
Xindong Wu
Lakhmi C. Jain
David Padua
Xuemin Shen
Borko Furht
VS Subrahmanian
Martial Hebert
Katsushi Ikeuchi
Bruno Siciliano

T0222302

For further volumes:
http://www.springer.com/series/10028

Maria Vanina Martinez • Cristian Molinaro
V.S. Subrahmanian • Leila Amgoud

A General Framework for Reasoning On Inconsistency

 Springer

Maria Vanina Martinez
Department of Computer Science
University of Oxford
Oxford, UK

V.S. Subrahmanian
Department of Computer Science
Inst. Advanced Computer Studies
University of Maryland
College Park, MD, USA

Cristian Molinaro
Dipartimento di Elettronica
Università della Calabria
Rende, Italy

Leila Amgoud
IRIT-UPS
Toulouse, France

ISSN 2191-5768   ISSN 2191-5776 (electronic)
ISBN 978-1-4614-6749-6   ISBN 978-1-4614-6750-2 (eBook)
DOI 10.1007/978-1-4614-6750-2
Springer New York Heidelberg Dordrecht London

Library of Congress Control Number: 2013932297

Printed on acid-free paper

Springer is part of Springer Science+Business Media (www.springer.com)

Acknowledgements

Some of the authors of this monograph may have been funded in part by AFOSR grant FA95500610405, ARO grants W911NF0910206, W911NF1160215, W911NF1110344, an ARO/Penn State MURI award, and ONR grant N000140910685. Maria Vanina Martinez is currently partially supported by the Engineering and Physical Sciences Research Council of the United Kingdom (EPSRC) grant EP/J008346/1 ("PrOQAW: Probabilistic Ontological Query Answering on the Web") and by a Google Research Award. Finally, the authors would also like to thank the anonymous reviewers who provided valuable feedback on earlier versions of this work; their comments and suggestions helped improve this manuscript.

Contents

Chapter 1
Introduction and Preliminary Concepts

1.1 Motivation

Inconsistency management has been intensely studied in various parts of AI, often in slightly disguised form (Gardenfors 1988; Pinkas and Loui 1992; Poole 1985; Rescher and Manor 1970). For example, default logics (Reiter 1980) use syntax to distinguish between strict facts and default rules, and identify different extensions of the default logic as potential ways of "making sense" of seemingly conflicting information. Likewise, inheritance networks (Touretzkey 1984) define extensions based on analyzing paths in the network and using notions of specificity to resolve conflicts. Argumentation frameworks (Dung 1995) study different ways in which an argument for or against a proposition can be made, and then determine which arguments defeat which other arguments in an effort to decide what can be reasonably concluded. All these excellent works provide an a priori conflict resolution mechanism. A user who uses a system based on these papers is forced to use the semantics implemented in the system, and has little say in the matter (besides which most users querying KBs are unlikely to be experts in even classical logic, let alone default logics and argumentation methods).

The aims of the monograph are:

1. To propose a unified framework for reasoning about inconsistency, which captures existing approaches as a special case and provides an easy basis for comparison;
2. To apply the framework using any monotonic logic, including ones for which inconsistency management has not been studied before (e.g., temporal, spatial, and probabilistic logics), and provide new results on the complexity of reasoning about inconsistency in such logics;
3. To allow end-users to bring their domain knowledge to bear, allowing them to voice an opinion on *what works for them, not what a system manager decided was right for them*, in other words, to take into account the preferences of the end-user;

4. To propose the concept of an *option* that specifies the semantics of an inconsistent theory in any of these monotonic logics and the notion of a preferred option that takes the user's domain knowledge into account; and
5. To propose general algorithms for computing the preferred options.

We do this by building upon Alfred Tarski and Dana Scott's celebrated notion of an abstract logic. We start with a simple example to illustrate why conflicts can often end up being resolved in different ways by human beings, and why it is important to allow end-users to bring their knowledge to bear when a system resolves conflicts. A database system designer or an AI knowledge base designer cannot claim to understand a priori the specifics of each application that his knowledge base system may be used for in the future.

Example 1.1. Suppose a university payroll system says that John's salary is 50K, while the university personnel database says it is 60K. In addition, there may be an axiom that says that everyone has exactly one salary. One simple way to model this is via the theory S below.

$$salary(John, 50K) \leftarrow \tag{1.1}$$

$$salary(John, 60K) \leftarrow \tag{1.2}$$

$$S_1 = S_2 \leftarrow salary(X, S_1) \wedge salary(X, S_2). \tag{1.3}$$

The above theory is obviously inconsistent. Suppose (1.3) is definitely known to be true. Then, a bank manager considering John for a loan may choose the 50K number to determine the maximum loan amount that John qualifies for. On the other hand, a national tax agency may use the 60K figure to send John a letter asking him why he underpaid his taxes.

Neither the bank manager nor the tax officer is making any attempt to find out the truth (thus far); however, both of them are making different decisions based on the same facts.

The following examples present theories expressed in different logics which are inconsistent – thus the reasoning that can be done is very limited. We will continue these examples later on to show how the proposed framework is suitable for handling all these scenarios in a flexible way.

Example 1.2. Consider the temporal logic theory T below. The $\bigcirc$ operator denotes the "next time instant" operator. Thus, the first rule says that if *received* is true at time t (intuitively, a request is received at time t), then *processed* is true at time $t + 1$ (the request is processed at the next time point).

$$\bigcirc processed \leftarrow received. \tag{1.4}$$

$$received. \tag{1.5}$$

$$\bigcirc \neg processed. \tag{1.6}$$

Clearly, the theory is inconsistent. Nevertheless, there might be several options that a user might consider to handle it: for instance, one might replace the first rule with a "weaker" one stating that if a request is received, it will be processed sometime in the future, not necessarily at the next time point.

Example 1.3. Consider the probabilistic logic theory P consisting of three formulas $p : [0.3, 0.4]$, $p : [0.41, 0.43]$, $p : [0.44, 0.6]$. In general, the formula $p : [\ell, u]$ says that the probability of proposition p lies in the interval $[\ell, u]$. The theory is easily seen to be inconsistent under this informal reading. A user might choose to resolve the inconsistency in different ways, such as by discarding a minimal set of formulas or modifying the probability intervals as little as possible.

The rest of this monograph proceeds as follows. In Sect. 1.2, we recall Tarski's notion of what an abstract logic is Tarski (1956). Then, in Chap. 2, we define our general framework for reasoning about inconsistency for any Tarskian logic. In Chap. 3, we develop general algorithms to compute preferred options based on various types of assumptions. In Chap. 4, we show applications of our framework to several monotonic logics for which no methods to reason about inconsistency exist to date – these logics include probabilistic logics (Halpern 1990; Nilsson 1986), linear temporal logic of the type used extensively in model checking (Emerson 1990; Gabbay et al. 1980), fuzzy logic, Levesque's logic of belief (Levesque 1984) and spatial logic captured via the region connection calculus (Randell et al. 1992). In many of these cases, we are able to establish new results on the complexity of reasoning about inconsistency in such logics. In Chap. 5, we show how certain existing works can be captured in our general framework.

1.2 Tarski's Abstract Logic

Alfred Tarski (1956) defines an *abstract logic* as a pair $(\mathscr{L}, \mathsf{CN})$ where the members of $\mathscr{L}$ are called *well-formed formulas*, and CN is a *consequence operator*. CN is any function from $2^{\mathscr{L}}$ (the powerset of $\mathscr{L}$) to $2^{\mathscr{L}}$ that satisfies the following axioms (here X is a subset of $\mathscr{L}$):

1. $X \subseteq \mathsf{CN}(X)$ (**Expansion**)
2. $\mathsf{CN}(\mathsf{CN}(X)) = \mathsf{CN}(X)$ (**Idempotence**)
3. $\mathsf{CN}(X) = \bigcup_{Y \subseteq_f X} \mathsf{CN}(Y)$ (**Finiteness**)
4. $\mathsf{CN}(\{x\}) = \mathscr{L}$ for some $x \in \mathscr{L}$ (**Absurdity**)

Notation: $Y \subseteq_f X$ means that Y is a finite subset of X.

Intuitively, $\mathsf{CN}(X)$ returns the set of formulas that are logical consequences of X according to the logic in question. It can be easily shown from the above axioms that CN is a closure operator, that is, for any $X, X', X'' \subseteq \mathscr{L}$, CN enjoys the following properties:

5. $X \subseteq X' \Rightarrow \mathsf{CN}(X) \subseteq \mathsf{CN}(X')$.   (**Monotonicity**)
6. $\mathsf{CN}(X) \cup \mathsf{CN}(X') \subseteq \mathsf{CN}(X \cup X')$.
7. $\mathsf{CN}(X) = \mathsf{CN}(X') \Rightarrow \mathsf{CN}(X \cup X'') = \mathsf{CN}(X' \cup X'')$.

Most well-known monotonic logics (such as first order logic (Shoenfield 1967), propositional logic (Shoenfield 1967), modal logic, temporal logic, fuzzy logic, probabilistic logic (Bacchus 1990), etc.) are special cases of Tarski's notion of an abstract logic. AI introduced non-monotonic logics (Bobrow 1980) which do not satisfy the monotonicity property.

Once $(\mathscr{L}, \text{CN})$ is fixed, a notion of *consistency* arises as follows.

Definition 1.1 (Consistency). Let $X \subseteq \mathscr{L}$. X is *consistent* w.r.t. the logic $(\mathscr{L}, \text{CN})$ iff $\text{CN}(X) \neq \mathscr{L}$. It is *inconsistent* otherwise.

The previous definition says that X is consistent iff its set of consequences is not the set of all formulas. For any abstract logic $(\mathscr{L}, \text{CN})$, we also require the following axiom to be satisfied:

8. $\text{CN}(\emptyset) \neq \mathscr{L}$ **(Coherence)**

The coherence requirement (absent from Tarski's original proposal, but added here to avoid considering trivial systems) forces the empty set $\emptyset$ to always be consistent – this makes sense for any reasonable logic as saying emptiness should intuitively be consistent.

It can be easily verified that if a set $X \subseteq \mathscr{L}$ is consistent, then its closure under CN is consistent as well as any subset of X. Moreover, if X is inconsistent, then every superset of X is inconsistent.

Chapter 2
A General Framework for Handling Inconsistency

This chapter proposes a general framework for handling inconsistency under any monotonic logic. Basically, our approach to reason with an inconsistent knowledge base (KB) is a process which follows three steps:

1. Determining consistent "subbases";
2. Selecting among all the subbases the ones that are *preferred*;
3. Applying entailment on the preferred subbases.

Throughout the rest of this chapter, we assume that we have an arbitrary, but fixed *monotonic* logic $(\mathscr{L}, \mathsf{CN})$.

The basic idea behind our framework is to construct what we call *options*, and then to define a preference relation on these options. The *preferred options* are intended to support the conclusions to be drawn from the inconsistent knowledge base. Intuitively, an option is a set of formulas that is both consistent and closed w.r.t. consequence in logic $(\mathscr{L}, \mathsf{CN})$.

Definition 2.1 (Options). An *option* is any set $\mathscr{O}$ of elements of $\mathscr{L}$ such that:

- $\mathscr{O}$ is consistent.
- $\mathscr{O}$ is closed, i.e., $\mathscr{O} = \mathsf{CN}(\mathscr{O})$.

We use $\mathsf{Opt}(\mathscr{L})$ to denote the set of all options that can be built from $(\mathscr{L}, \mathsf{CN})$.

Note that the empty set is not necessarily an option. This depends on the value of $\mathsf{CN}(\emptyset)$ in the considered logic $(\mathscr{L}, \mathsf{CN})$. For instance, in propositional logic, it is clear that $\emptyset$ is not an option since all the tautologies will be inferred from it. Indeed, it is easy to see that $\emptyset$ is an option iff $\mathsf{CN}(\emptyset) = \emptyset$.

Clearly, for each consistent subset X of $\mathscr{L}$, it holds that $\mathsf{CN}(X)$ is an option (as $\mathsf{CN}(X)$ is consistent and Idempotence axiom entails that $\mathsf{CN}(X)$ is closed). Since we are considering generic logic, we can show that options do not always exist.

Proposition 2.1. *The set* $\mathsf{Opt}(\mathscr{L}) = \emptyset$ *iff*

1. *$\forall \psi \in \mathscr{L}$, $\mathsf{CN}(\{\psi\})$ is inconsistent, and*
2. *$\mathsf{CN}(\emptyset) \neq \emptyset$.*

Proof. ($\Rightarrow$) Let us assume that $\mathrm{Opt}(\mathscr{L}) = \emptyset$. Let us also assume by contradiction that $\exists \psi \in \mathscr{L}$ such that $\mathrm{CN}(\{\psi\})$ is consistent. Since $\mathrm{CN}(\{\psi\})$ is closed by the Idempotence axiom, $\mathrm{CN}(\{\psi\})$ is an option, which is a contradiction.

Assume now that $\mathrm{CN}(\emptyset) = \emptyset$. This means that $\emptyset$ is an option since it is closed and consistent ($\mathrm{CN}(\emptyset) \neq \mathscr{L}$), which is a contradiction.

($\Leftarrow$) Let us assume that (i) $\forall \psi \in \mathscr{L}$, $\mathrm{CN}(\{\psi\})$ is inconsistent, and (ii) $\mathrm{CN}(\emptyset) \neq \emptyset$. Assume also by contradiction that $\mathrm{Opt}(\mathscr{L}) \neq \emptyset$ and let $\mathscr{O} \in \mathrm{Opt}(\mathscr{L})$. There are two cases:

Case 1: $\mathscr{O} = \emptyset$. Consequently, $\mathrm{CN}(\emptyset) = \emptyset$, which contradicts assumption (ii).
Case 2: $\mathscr{O} \neq \emptyset$. Since $\mathscr{O}$ is consistent, $\exists \psi \in \mathscr{O}$ s.t. $\{\psi\}$ is consistent and thus $\mathrm{CN}(\{\psi\})$ is consistent. This contradicts assumption (i). $\square$

So far, we have defined the concept of option for any logic $(\mathscr{L}, \mathrm{CN})$ in a way that is independent of a knowledge base. We now show how to associate a set of options with an inconsistent knowledge base.

In most approaches for handling inconsistency, the maximal consistent subsets of a given inconsistent knowledge base have an important role. This may induce one to think of determining the options of a knowledge base as the closure of its maximal consistent subsets. However, this approach has the side effect of dropping entire formulas, whereas more fine-grained approaches could be adopted in order to preserve more information of the original knowledge base. This is shown in the following example.

Example 2.1. Consider the propositional knowledge base $\mathscr{K} = \{(a \wedge b); \neg b\}$. There are two maximal consistent subsets, namely $MCS_1 = \{a \wedge b\}$ and $MCS_2 = \{\neg b\}$. However, one could argue that MCS_2 is too weak, since we could have included a by "weakening" the formula $(a \wedge b)$ instead of dropping it altogether.

The "maximal consistent subset" approach, as well as the one suggested in the previous example, can be seen as a particular case of a more general approach, where one considers consistent "relaxations" (or *weakenings*) of a given inconsistent knowledge base. The ways in which such weakenings are determined might be different, as the following examples show.

Example 2.2. Consider again the temporal knowledge base of Example 1.2. An intuitive way to "weaken" the knowledge base might consist of replacing the $\bigcirc$ (*next moment in time*) connective with the $\Diamond$ (*sometime in the future*) connective. So, for instance, $\bigcirc processed \leftarrow received$ might be replaced by $\Diamond processed \leftarrow received$, thus saying that if *received* is true at time t, then *processed* is true at some subsequent time $t' \geq t$ (not necessarily at time $t+1$). This would lead to a consistent knowledge base, whose closure is clearly an option. Likewise, we might weaken only (1.6), obtaining another consistent knowledge base whose closure is an option.

Example 2.3. Consider the probabilistic knowledge base of Example 1.3. A reasonable way to make a probabilistic formula $\phi : [\ell, u]$ weaker, might be to replace it with another formula $\phi : [\ell', u']$ where $[\ell, u] \subseteq [\ell', u']$.

The preceding examples suggest that a flexible way to determine the options of a given knowledge base should be provided, since what is considered reasonable to be an option might depend on the logic and the application domain at hand, and, more importantly, it should depend on the user's preferences. The basic idea is to consider *weakenings* of a given knowledge base $\mathcal{K}$ whose closures yield options. For instance, as said before, weakenings might be subsets of the knowledge base. Although such a weakening mechanism is general enough to be applicable to many logics, more tailored mechanisms could be defined for specific logics. For instance, the two reasonable approaches illustrated in Examples 2.2 and 2.3 above cannot be captured by considering subsets of the original knowledge bases; as another example, let us reconsider Example 2.1: by looking at subsets of the knowledge base, it is not possible to get an option containing both a and $\neg b$. We formally introduce the notion of *weakening* as follows.

Definition 2.2. Given an element ψ of $\mathcal{L}$,

$$weakening(\psi) = \begin{cases} \mathsf{CN}(\{\psi\}) & if\ \psi\ is\ consistent \\ \emptyset & otherwise \end{cases}$$

Definition 2.3. Given a knowledge base $\mathcal{K}$,

$$weakening(\mathcal{K}) = \{\mathcal{K}' \subseteq \mathcal{L} \mid \forall \psi' \in \mathcal{K}'\ (\exists \psi \in \mathcal{K}.\ \psi' \in weakening(\psi))\}$$

According to the preceding definitions, to weaken a knowledge base intuitively means to weaken formulas in it; to weaken a formula ψ means to take some formulas in $\mathsf{CN}(\{\psi\})$ if ψ is consistent, or to otherwise drop ψ altogether (note that a consistent formula could also be dropped). The set $weakening(\mathcal{K})$ can be computed by first finding $weakening(\psi)$ for all $\psi \in \mathcal{K}$ and then returning the subsets of $\bigcup_{\psi \in \mathcal{K}} weakening(\psi)$. It is easy to see that if $\mathcal{K}' \in weakening(\mathcal{K})$, then $\mathsf{CN}(\mathcal{K}') \subseteq \mathsf{CN}(\mathcal{K})$.

Observe that although a knowledge base in $weakening(\mathcal{K})$ does not contain any inconsistent formulas, it could be inconsistent.

Definition 2.4. A *weakening mechanism* is a function $\mathcal{W} : 2^{\mathcal{L}} \rightarrow 2^{2^{\mathcal{L}}}$ such that $\mathcal{W}(\mathcal{K}) \subseteq weakening(\mathcal{K})$ for any $\mathcal{K} \in 2^{\mathcal{L}}$.

The preceding definition says that a weakening mechanism is a function that maps a knowledge base into knowledge bases that are *weaker* in some sense. For instance, an example of a weakening mechanism is $\mathcal{W}(\mathcal{K}) = weakening(\mathcal{K})$. This returns *all* the weaker knowledge bases associated with $\mathcal{K}$. We use $\mathcal{W}_{all}$ to denote this weakening mechanism.

We now define the set of options for a given knowledge base (w.r.t. a selected weakening mechanism).

Definition 2.5. Let $\mathcal{K}$ be a knowledge base in logic $(\mathcal{L}, \mathsf{CN})$ and $\mathcal{W}$ a weakening mechanism. We say that an option $\mathcal{O} \in \mathsf{Opt}(\mathcal{L})$ is an option for $\mathcal{K}$ (w.r.t. $\mathcal{W}$) iff there exists $\mathcal{K}'$ in $\mathcal{W}(\mathcal{K})$ such that $\mathcal{O} = \mathsf{CN}(\mathcal{K}')$.

Thus, an option for $\mathcal{K}$ is the closure of some weakening $\mathcal{K}'$ of $\mathcal{K}$. Clearly, $\mathcal{K}'$ must be consistent because $\mathcal{O}$ is consistent (by virtue of being an option) and because $\mathcal{O} = \mathsf{CN}(\mathcal{K}')$. In other words, the options for $\mathcal{K}$ are the closure of consistent weakenings of $\mathcal{K}$. We use $\mathsf{Opt}(\mathcal{K}, \mathcal{W})$ to denote the set of options for $\mathcal{K}$ under the weakening mechanism $\mathcal{W}$. Whenever $\mathcal{W}$ is clear from the context, we simply write $\mathsf{Opt}(\mathcal{K})$ instead of $\mathsf{Opt}(\mathcal{K}, \mathcal{W})$.

Note that if we restrict $\mathcal{W}(\mathcal{K})$ to be $\{\mathcal{K}' \mid \mathcal{K}' \subseteq \mathcal{K}\}$, Definition 2.5 corresponds to that presented in Subrahmanian and Amgoud (2007) (we will refer to such a weakening mechanism as $\mathcal{W}_\subseteq$). Moreover, observe that every option for a knowledge base w.r.t. this weakening mechanism is also an option for the knowledge base when $\mathcal{W}_{all}$ is adopted, that is, the options obtained in the former case are a subset of those obtained in the latter case.

Example 2.4. Consider again the knowledge base of Example 2.1 and let $\mathcal{W}_{all}$ be the adopted weakening mechanism. Our framework is flexible enough to allow the set $\mathsf{CN}(\{a, \neg b\})$ to be an option for $\mathcal{K}$. This weakening mechanism preserves more information from the original knowledge base than the classical "maximal consistent subsets" approach.

In Chap. 4 we will consider specific monotonic logics and show more tailored weakening mechanisms.

The framework for reasoning about inconsistency has three components: the set of all options for a given knowledge base, a preference relation between options, and an inference mechanism.

Definition 2.6 (General framework). A *general framework for reasoning about inconsistency* in a knowledge base $\mathcal{K}$ is a triple $\langle \mathsf{Opt}(\mathcal{K}, \mathcal{W}), \succeq, \sim \rangle$ such that:

- $\mathsf{Opt}(\mathcal{K}, \mathcal{W})$ is the set of options for $\mathcal{K}$ w.r.t. the weakening mechanism $\mathcal{W}$.
- $\succeq \subseteq \mathsf{Opt}(\mathcal{K}, \mathcal{W}) \times \mathsf{Opt}(\mathcal{K}, \mathcal{W})$. $\succeq$ is a partial (or total) preorder (i.e., it is reflexive and transitive).
- $\sim : 2^{\mathsf{Opt}(\mathcal{K}, \mathcal{W})} \to \mathsf{Opt}(\mathcal{L})$.

The second important concept of the general framework above is the preference relation $\succeq$ among options. Indeed, $\mathcal{O}_1 \succeq \mathcal{O}_2$ means that the option $\mathcal{O}_1$ is at least as preferred as $\mathcal{O}_2$. This relation captures the idea that some options are better than others because, for instance, the user has decided that this is the case, or because those preferred options satisfy the requirements imposed by the developer of a conflict management system. For instance, in Example 1.1, the user chooses certain options (e.g., the options where the salary is minimal or where the salary is maximal based on his needs). From the partial preorder $\succeq$ we can derive the strict partial order $\succ$ (i.e., it is irreflexive and transitive) over $\mathsf{Opt}(\mathcal{K}, \mathcal{W})$ as follows: for any $\mathcal{O}_1, \mathcal{O}_2 \in \mathsf{Opt}(\mathcal{K}, \mathcal{W})$ we say $\mathcal{O}_1 \succ \mathcal{O}_2$ iff $\mathcal{O}_1 \succeq \mathcal{O}_2$ and $\mathcal{O}_2 \not\succeq \mathcal{O}_1$. Intuitively, $\mathcal{O}_1 \succ \mathcal{O}_2$ means that $\mathcal{O}_1$ is *strictly* preferable to $\mathcal{O}_2$. The set of *preferred* options in $\mathsf{Opt}(\mathcal{K}, \mathcal{W})$ determined by $\succeq$ is $\mathsf{Opt}^\succeq(\mathcal{K}, \mathcal{W}) = \{\mathcal{O} \mid \mathcal{O} \in \mathsf{Opt}(\mathcal{K}, \mathcal{W}) \wedge \nexists \mathcal{O}' \in \mathsf{Opt}(\mathcal{K}, \mathcal{W}) \text{ with } \mathcal{O}' \succ \mathcal{O}\}$. Whenever $\mathcal{W}$ is clear from the context, we simply write $\mathsf{Opt}^\succeq(\mathcal{K})$ instead of $\mathsf{Opt}^\succeq(\mathcal{K}, \mathcal{W})$.

In the following three examples, we come back to the example theories of Chap. 1 to show how our framework can handle them.

Example 2.5. Let us consider again the knowledge base S of Example 1.1. Consider the options $\mathcal{O}_1 = \text{CN}(\{(1.1),(1.3)\})$, $\mathcal{O}_2 = \text{CN}(\{(1.1),(1.2)\})$, $\mathcal{O}_3 = \text{CN}(\{(1.2), (1.3)\})$, and let us say that these three options are strictly preferable to all other options for S; then, we have to determine the preferred options among these three. Different criteria might be used to determine the preferred options:

- Suppose the *score* $sc(\mathcal{O}_i)$ of option $\mathcal{O}_i$ is the sum of the elements in the multiset $\{S \mid salary(John, S) \in \mathcal{O}_i\}$. In this case, the score of $\mathcal{O}_1$ is 50K, that of $\mathcal{O}_2$ is 110K, and that of $\mathcal{O}_3$ is 60K. We could now say that $\mathcal{O}_i \succeq \mathcal{O}_j$ iff $sc(\mathcal{O}_i) \leq sc(\mathcal{O}_j)$. In this case, the only preferred option is $\mathcal{O}_1$, which corresponds to the bank manager's viewpoint.
- On the other hand, suppose we say that $\mathcal{O}_i \succeq \mathcal{O}_j$ iff $sc(\mathcal{O}_i) \geq sc(\mathcal{O}_j)$. In this case, the only preferred option is $\mathcal{O}_2$; this corresponds to the view that the rule saying everyone has only one salary is wrong (perhaps the database has John being paid out of two projects simultaneously and 50K of his salary is charged to one project and 60K to another).
- Now consider the case where we change our scoring method and say that $sc(\mathcal{O}_i) = \min\{S \mid salary(John, S) \in \mathcal{O}_i\}$. In this case, $sc(\mathcal{O}_1) = 50K$, $sc(\mathcal{O}_2) = 50K, sc(\mathcal{O}_3) = 60K$. Let us suppose that the preference relation says that $\mathcal{O}_i \succeq \mathcal{O}_j$ iff $sc(\mathcal{O}_i) \geq sc(\mathcal{O}_j)$. Then, the only preferred option is $\mathcal{O}_3$, which corresponds exactly to the tax agency's viewpoint.

Example 2.6. Let us consider the temporal logic theory T of Example 1.2. We may choose to consider just three options for determining the preferred ones: $\mathcal{O}_1 = \text{CN}(\{(1.4),(1.5)\})$, $\mathcal{O}_2 = \text{CN}(\{(1.4),(1.6)\})$, $\mathcal{O}_3 = \text{CN}(\{(1.5),(1.6)\})$. Suppose now that we can associate a numeric score with each formula in T, describing the reliability of the source that provided the formula. Let us say these scores are 3, 1, and 2 for formulas (1.4), (1.5) and (1.6), respectively, and the weight of an option $\mathcal{O}_i$ is the sum of the scores of the formulas in $T \cap \mathcal{O}_i$. We might say $\mathcal{O}_i \succeq \mathcal{O}_j$ iff the score of $\mathcal{O}_i$ is greater than or equal to the score of $\mathcal{O}_j$. In this case, the only preferred option is $\mathcal{O}_2$.

Example 2.7. Consider the probabilistic logic theory P of Example 1.3. Suppose that in order to determine the preferred options, we consider only options that assign a single non-empty probability interval to p, namely options of the form $\text{CN}(\{p : [\ell, u]\})$. For two atoms $A_1 = p : [\ell_1, u_1]$ and $A_2 = p : [\ell_2, u_2]$, let $diff(A_1, A_2) = abs(\ell_1 - \ell_2) + abs(u_1 - u_2)$. Let us say that the score of an option $\mathcal{O} = \text{CN}(\{A\})$, denoted by $score(\mathcal{O})$, is given by $\sum_{A' \in P} diff(A, A')$. Suppose we say that $\mathcal{O}_i \succeq \mathcal{O}_j$ iff $score(\mathcal{O}_i) \leq score(\mathcal{O}_j)$. Intuitively, this means that we are preferring options that change the lower and upper bounds in P as little as possible. In this case, $\text{CN}(\{p : [0.41, 0.43]\})$ is a preferred option.

Thus, we see that our general framework for managing inconsistency is very powerful – it can be used to handle inconsistencies in different ways based upon how the preference relation between options is defined. In Chap. 4, we will consider more logics and illustrate more examples showing how the proposed framework is suitable for handling inconsistency in a flexible way.

The following definition introduces a preference criterion where an option is preferable to another if and only if the latter is a weakening of the former.

Definition 2.7. Consider a knowledge base $\mathcal{K}$ and a weakening mechanism $\mathcal{W}$. Let $\mathcal{O}_1, \mathcal{O}_2 \in \text{Opt}(\mathcal{K}, \mathcal{W})$. We say $\mathcal{O}_1 \succeq_W \mathcal{O}_2$ iff $\mathcal{O}_2 \in weakening(\mathcal{O}_1)$.

Proposition 2.2. *Consider a knowledge base $\mathcal{K}$ and a weakening mechanism $\mathcal{W}$. Let $\mathcal{O}_1, \mathcal{O}_2 \in \text{Opt}(\mathcal{K}, \mathcal{W})$. $\mathcal{O}_1 \succeq_W \mathcal{O}_2$ iff $\mathcal{O}_1 \supseteq \mathcal{O}_2$.*

Proof. ($\Rightarrow$) Let $\psi_2 \in \mathcal{O}_2$. By definition of $\succeq_W$, there exists $\psi_1 \in \mathcal{O}_1$ s.t. $\psi_2 \in weakening(\psi_1)$; that is $\psi_2 \in \text{CN}(\{\psi_1\})$. Since $\{\psi_1\} \subseteq \mathcal{O}_1$, it follows that $\text{CN}(\{\psi_1\}) \subseteq \mathcal{O}_1$ (by Monotonicity and the fact that $\mathcal{O}_1$ is closed). Hence, $\psi_2 \in \mathcal{O}_1$.

($\Leftarrow$) Let $\psi_2 \in \mathcal{O}_2$. Clearly, $\psi_2 \in weakening(\psi_2)$, since ψ_2 is consistent and $\psi_2 \in \text{CN}(\{\psi_2\})$ (Expansion axiom). As $\psi_2 \in \mathcal{O}_1$, the condition expressed in Definition 2.3 trivially holds and $\mathcal{O}_1 \succeq_W \mathcal{O}_2$. $\square$

The following corollary states that $\succeq_W$ is indeed a preorder (in particular, a partial order).

Corollary 2.1. *Consider a knowledge base $\mathcal{K}$ and a weakening mechanism $\mathcal{W}$. $\succeq_W$ is a partial order over $\text{Opt}(\mathcal{K}, \mathcal{W})$.*

Proof. Straightforward from Proposition 2.2. $\square$

If the user's preferences are expressed according to $\succeq_W$, then the preferred options are the least weak or, in other words, in view of Proposition 2.2, they are the maximal ones under set inclusion.

The third component of the framework is a mechanism for selecting the inferences to be drawn from the knowledge base. In our framework, the set of inferences is itself an option. Thus, it should be consistent. This requirement is of great importance, since it ensures that the framework delivers *safe* conclusions. Note that this inference mechanism returns an option of the language from the set of options for a given knowledge base. The set of inferences is generally computed from the preferred options. Different mechanisms can be defined for selecting the inferences to be drawn. Here is an example of such a mechanism.

Definition 2.8 (Universal Consequences). Let $\langle \text{Opt}(\mathcal{K}, \mathcal{W}), \succeq, \mid\sim \rangle$ be a framework. A formula $\psi \in \mathcal{L}$ is a universal consequence of $\mathcal{K}$ iff $(\forall \mathcal{O} \in \text{Opt}^{\succeq}(\mathcal{K}, \mathcal{W}))$ $\psi \in \mathcal{O}$.

We can show that the set of inferences made using the universal criterion is itself an option of $\mathcal{K}$, and thus the universal criterion is a valid mechanism of inference. Moreover, it is included in every preferred option.

Proposition 2.3. *Let* $\langle \mathtt{Opt}(\mathcal{K}, \mathcal{W}), \succeq, \hspace{-0.3em}\sim \rangle$ *be a framework. The set* $\{\psi \mid \psi$ *is a universal consequence of* $\mathcal{K}\}$ *is an option in* $\mathtt{Opt}(\mathcal{L})$.

Proof. Let $\mathcal{C} = \{\psi \mid \psi$ is a universal consequence of $\mathcal{K}\}$. As each $\mathcal{O}_i \in \mathtt{Opt}^{\succeq}$ $(\mathcal{K}, \mathcal{W})$ is an option, $\mathcal{O}_i$ is consistent. Thus, $\mathcal{C}$ (which is a subset of every $\mathcal{O}_i$) is also consistent. Moreover, since $\mathcal{C} \subseteq \mathcal{O}_i$, thus $\mathtt{CN}(\mathcal{C}) \subseteq \mathcal{O}_i$ (by Monotonicity and Idempotence axioms), $\forall \mathcal{O}_i \in \mathtt{Opt}^{\succeq}(\mathcal{K}, \mathcal{W})$. Consequently, $\mathtt{CN}(\mathcal{C}) \subseteq \mathcal{C}$ (according to the above definition of universal consequences). In particular, $\mathtt{CN}(\mathcal{C}) = \mathcal{C}$ because of the expansion axiom. Thus, $\mathcal{C}$ is closed and consistent, and is therefore an option in $\mathtt{Opt}(\mathcal{L})$. $\square$

However, the following criterion

$$\mathcal{K} \hspace{0.2em}\sim \psi \text{ iff } \exists \mathcal{O} \in \mathtt{Opt}^{\succeq}(\mathcal{K}, \mathcal{W}) \text{ such that } \psi \in \mathcal{O}$$

is not a valid inference mechanism since the set of consequences returned by it may be inconsistent, thus, it is not an option.

Chapter 3
Algorithms

In this chapter, we present now general algorithms for computing the preferred options for a given knowledge base. Throughout this chapter, we assume that $CN(\mathcal{K})$ is finite for any knowledge base $\mathcal{K}$. The preferred options could be naively computed as follows.

procedure CPO-Naive$(\mathcal{K}, \mathcal{W}, \succeq)$
1.   Let $X = \{CN(\mathcal{K}') \mid \mathcal{K}' \in \mathcal{W}(\mathcal{K}) \wedge \mathcal{K}'$ is consistent$\}$
2.   Return any $\mathcal{O} \in X$ s.t. there is no $\mathcal{O}' \in X$ s.t. $\mathcal{O}' \succ \mathcal{O}$

Clearly, X is the set of options for $\mathcal{K}$. Among them, the algorithm chooses the preferred ones according to $\succeq$. Note that **CPO-Naive**, as well as the other algorithms we present in the following, relies on the CN operator, which makes the algorithm independent of the underlying logic; in order to apply the algorithm to a specific logic it suffices to provide the definition of CN for that logic. One reason for the inefficiency of **CPO-Naive** is that it makes no assumptions about the weakening mechanism and the preference relation.

The next theorem identifies the set of preferred options for a given knowledge base when $\mathcal{W}_{all}$ and $\succeq_W$ are the weakening mechanism and the preference relation, respectively.

Theorem 3.1. *Consider a knowledge base $\mathcal{K}$. Let $\mathcal{W}_{all}$ and $\succeq_W$ be the weakening mechanism and preference relation, respectively, that are used. Let $\Phi = \bigcup_{\psi \in \mathcal{K}} weakening(\psi)$. Then, the set of preferred options for $\mathcal{K}$ is equal to $\mathcal{PO}$ where*

$$\mathcal{PO} = \{CN(\mathcal{K}') \mid \mathcal{K}' \text{ is a maximal consistent subset of } \Phi\}$$

Proof. First, we show that any $\mathcal{O} \in \mathcal{PO}$ is a preferred option for $\mathcal{K}$. Let $\mathcal{K}'$ be a maximal consistent subset of Φ s.t. $\mathcal{O} = CN(\mathcal{K}')$. It is easy to see that $\mathcal{K}'$ is in $\mathcal{W}_{all}(\mathcal{K})$. Since $\mathcal{K}'$ is consistent and $\mathcal{O} = CN(\mathcal{K}')$, then $\mathcal{O}$ is an option for $\mathcal{K}$. Suppose by contradiction that $\mathcal{O}$ is not preferred, i.e., there exists an option $\mathcal{O}'$ for $\mathcal{K}$ s.t. $\mathcal{O}' \succ \mathcal{O}$. Proposition 2.2 entails that $\mathcal{O}' \supset \mathcal{O}$. Since $\mathcal{O}'$ is an option for $\mathcal{K}$,

M.V. Martinez et al., *A General Framework for Reasoning On Inconsistency*, SpringerBriefs
in Computer Science, DOI 10.1007/978-1-4614-6750-2_3, © The Author(s) 2013

then there exists a weakening $\mathscr{W}' \in \mathscr{W}_{all}(\mathscr{K})$ s.t. $\mathcal{O}' = \mathrm{CN}(\mathscr{W}')$. There must be a formula $\psi' \in \mathscr{W}'$ which is not in $\mathcal{O}$ (hence $\psi' \notin \mathscr{K}'$), otherwise it would be the case that $\mathscr{W}' \subseteq \mathcal{O}$ and thus $\mathrm{CN}(\mathscr{W}') \subseteq \mathcal{O}$ (from Monotonicity and Idempotence axioms), that is $\mathcal{O}' \subseteq \mathcal{O}$. Since ψ' is in a weakening of $\mathscr{K}$, then there is a (consistent) formula $\psi \in \mathscr{K}$ s.t. $\psi' \in weakening(\psi)$ and therefore $\psi' \in \Phi$. As $\mathscr{K}' \subseteq \mathcal{O} \subset \mathcal{O}'$ and $\psi' \in \mathcal{O}'$, then $\mathscr{K}' \cup \{\psi'\}$ is consistent. Since $\psi' \notin \mathscr{K}'$, $\psi' \in \Phi$, and $\mathscr{K}' \cup \{\psi'\}$ is consistent, then $\mathscr{K}'$ is not a maximal consistent subset of Φ, which is a contradiction.

We now show that every preferred option $\mathcal{O}$ for $\mathscr{K}$ is in $\mathscr{P}\mathcal{O}$. Let $\mathscr{W}$ be a (consistent) weakening of $\mathscr{K}$ s.t. $\mathrm{CN}(\mathscr{W}) = \mathcal{O}$. It is easy to see that $\mathscr{W} \subseteq \Phi$. Then, there is a maximal consistent subset $\mathscr{K}'$ of Φ s.t. $\mathscr{W} \subseteq \mathscr{K}'$. Clearly, $\mathcal{O}' = \mathrm{CN}(\mathscr{K}')$ is in $\mathscr{P}\mathcal{O}$, and thus, as shown above, it is a preferred option for $\mathscr{K}$. Monotonicity entails that $\mathrm{CN}(\mathscr{W}) \subseteq \mathrm{CN}(\mathscr{K}')$, that is $\mathcal{O} \subseteq \mathcal{O}'$. In particular, $\mathcal{O} = \mathcal{O}'$, otherwise Proposition 2.2 would entail that $\mathcal{O}$ is not preferred. $\qquad\square$

Example 3.1. Consider again the knowledge base $\mathscr{K} = \{(a \wedge b); \neg b\}$ of Example 2.1. We have that $\Phi = \mathrm{CN}(\{a \wedge b\}) \cup \mathrm{CN}(\{\neg b\})$. Thus, it is easy to see that a preferred option for $\mathscr{K}$ is $\mathrm{CN}(\{a, \neg b\})$ (note that $a \in \Phi$ since $a \in \mathrm{CN}(\{a \wedge b\})$).

Clearly, we can straightforwardly derive an algorithm to compute the preferred options from the theorem above: first Φ is computed and then CN is applied to the maximal consistent subsets of Φ. Thus, such an algorithm does not need to compute all the options for a given knowledge base in order to determine the preferred ones (which is the case in the **CPO-Naive** algorithm) as every option computed by the algorithm is ensured to be preferred.

Example 3.2. Consider the following inconsistent propositional Horn[1] knowledge base $\mathscr{K}$:

$$a_1$$
$$a_2 \leftarrow a_1$$
$$a_3 \leftarrow a_2$$
$$\vdots$$
$$a_{n-1} \leftarrow a_{n-2}$$
$$\neg a_1 \leftarrow a_{n-1}$$

Suppose we want to compute one preferred option for $\mathscr{K}$ ($\mathscr{W}_{all}$ and $\succeq_W$ are the weakening mechanism and preference relation, respectively). If we use Algorithm **CPO-Naive**, then all the options for $\mathscr{K}$ w.r.t. $\mathscr{W}_{all}$ need to be computed in order to determine a preferred one. Observe that the closure of a proper subset of $\mathscr{K}$ is an option for $\mathscr{K}$, and thus the number of options is exponential. According to Theorem 3.1, a preferred option may be computed as $\mathrm{CN}(\mathscr{K}')$, where $\mathscr{K}'$ is a maximal consistent subset of $\bigcup_{\psi \in \mathscr{K}} weakening(\psi)$.

---

[1] Recall that a Horn clause is a disjunction of literals containing at most one positive literal.

Note that Theorem 3.1 entails that if both computing CN and consistency checking can be done in polynomial time, then one preferred option can be computed in polynomial time. For instance, this is the case for propositional Horn knowledge bases (see Chap. 4). Furthermore, observe that Theorem 3.1 also holds when $\supseteq$ is the preference relation simply because $\supseteq$ coincides with $\succeq_W$ (see Proposition 2.2).

Let us now consider the case where $\mathcal{W}_\subseteq$ and $\supseteq$ are the adopted weakening mechanism and preference relation, respectively.

Theorem 3.2. *Consider a knowledge base $\mathcal{K}$. Let $\mathcal{W}_\subseteq$ and $\supseteq$ respectively be the weakening mechanism and preference relation used. Then, a knowledge base $\mathcal{O}$ is a preferred option for $\mathcal{K}$ iff $\mathcal{K}' = \mathcal{O} \cap \mathcal{K}$ is a maximal consistent subset of $\mathcal{K}$ and $CN(\mathcal{K}') = \mathcal{O}$.*

Proof. ($\Leftarrow$) Clearly, $\mathcal{O}$ is an option for $\mathcal{K}$. Suppose by contradiction that $\mathcal{O}$ is not preferred, i.e., there exists an option $\mathcal{O}'$ for $\mathcal{K}$ s.t. $\mathcal{O} \subset \mathcal{O}'$. Since $\mathcal{O}'$ is an option for $\mathcal{K}$, then there exists $\mathcal{W} \subseteq \mathcal{K}$ s.t. $\mathcal{O}' = CN(\mathcal{W})$. There must be a formula $\psi \in \mathcal{W}$ which is not in $\mathcal{O}$ (hence $\psi \notin \mathcal{K}'$), otherwise it would be the case that $\mathcal{W} \subseteq \mathcal{O}$ and thus $CN(\mathcal{W}) \subseteq \mathcal{O}$ (from Monotonicity and Idempotence axioms), that is $\mathcal{O}' \subseteq \mathcal{O}$. As $\mathcal{K}' \subseteq \mathcal{O} \subset \mathcal{O}'$ and $\psi \in \mathcal{O}'$, then $\mathcal{K}' \cup \{\psi\}$ is consistent, that is $\mathcal{K}'$ is not a maximal consistent subset of $\mathcal{K}$, which is a contradiction.

($\Rightarrow$) Suppose by contradiction that $\mathcal{O}$ is a preferred option for $\mathcal{K}$ and a case of the following occurs: (i) $CN(\mathcal{K}') \neq \mathcal{O}$, (ii) $\mathcal{K}'$ is not a maximal consistent subset of $\mathcal{K}$.

(i) Since $\mathcal{K}' \subseteq \mathcal{O}$, then $CN(\mathcal{K}') \subseteq \mathcal{O}$ (Monotonicity and Idempotence axioms). As $CN(\mathcal{K}') \neq \mathcal{O}$, then $CN(\mathcal{K}') \subset \mathcal{O}$. Since $\mathcal{O}$ is an option, then there exists $\mathcal{W} \subseteq \mathcal{K}$ s.t. $\mathcal{O} = CN(\mathcal{W})$. Two cases may occur:

-   $\mathcal{W} \subseteq \mathcal{K}'$. Thus, $CN(\mathcal{W}) \subseteq CN(\mathcal{K}')$ (Monotonicity), i.e., $\mathcal{O} \subseteq CN(\mathcal{K}')$, which is a contradiction.
-   $\mathcal{W} \not\subseteq \mathcal{K}'$. Thus, there exists a formula ψ which is in $\mathcal{W}$ but not in $\mathcal{K}'$. Note that $\psi \in \mathcal{K}$ (as $\mathcal{W} \subseteq \mathcal{K}$) and $\psi \in \mathcal{O}$ (from the fact that $\mathcal{O} = CN(\mathcal{W})$ and Expansion axiom). Since $\mathcal{K}' = \mathcal{K} \cap \mathcal{O}$, then $\psi \in \mathcal{K}'$, which is a contradiction.

(ii) Since $\mathcal{K}' \subseteq \mathcal{O}$ then $\mathcal{K}'$ is consistent and is not maximal. Thus, there exists $\mathcal{K}'' \subseteq \mathcal{K}$ which is consistent and $\mathcal{K}' \subset \mathcal{K}''$. Monotonicity implies that $CN(\mathcal{K}') \subseteq CN(\mathcal{K}'')$, i.e., $\mathcal{O} \subseteq CN(\mathcal{K}'')$ since we have proved before that $\mathcal{O} = CN(\mathcal{K}')$. Let $\psi \in \mathcal{K}'' - \mathcal{K}'$. Since $\psi \in \mathcal{K}$ (as $\mathcal{K}'' \subseteq \mathcal{K}$) and $\psi \notin \mathcal{K}'$, then $\psi \notin \mathcal{O}$ (because $\mathcal{K}' = \mathcal{O} \cap \mathcal{K}$). Thus, $\mathcal{O} \subset CN(\mathcal{K}'')$. Since $\mathcal{K}''$ is consistent, then $CN(\mathcal{K}'')$ is an option and $\mathcal{O}$ is not preferred, which is a contradiction. □

The following corollary identifies the set of preferred options for a knowledge base when the weakening mechanism and the preference relation are $\mathcal{W}_\subseteq$ and $\supseteq$, respectively.

Corollary 3.1. *Consider a knowledge base* $\mathcal{K}$. *Let* $\mathcal{W}_{\subseteq}$ *and* $\supseteq$ *be the employed weakening mechanism and preference relation, respectively. Then, the set of preferred options for* $\mathcal{K}$ *is:*

$$\{CN(\mathcal{K}') \mid \mathcal{K}' \text{ is a maximal consistent subset of } \mathcal{K}\}$$

Proof. Straightforward from Theorem 3.2.   □

The preceding corollary provides a way to compute the preferred options: first the maximal consistent subsets of $\mathcal{K}$ are computed, then CN is applied to them. Clearly, such an algorithm avoids the computation of every option. Note that this corollary entails that if both computing CN and consistency checking can be done in polynomial time, then one preferred option can be computed in polynomial time. Moreover, observe that both the corollary above and Theorem 3.2 also hold in the case where the adopted preference criterion is $\succeq_W$ because $\supseteq$ coincides with $\succeq_W$ (see Proposition 2.2).

We now consider the case where different assumptions on the preference relation are made. The algorithms below are independent of the weakening mechanism that we choose to use. For the sake of simplicity, we will use $\text{Opt}(\mathcal{K})$ instead of $\text{Opt}(\mathcal{K}, \mathcal{W})$ to denote the set of options for a knowledge base $\mathcal{K}$.

Definition 3.1. A preference relation $\succeq$ is said to be *monotonic* iff for any $X, Y \subseteq \mathcal{L}$, if $X \subseteq Y$, then $Y \succeq X$. $\succeq$ is said to be *anti-monotonic* iff for any $X, Y \subseteq \mathcal{L}$, if $X \subseteq Y$, then $X \succeq Y$.

We now define the set of minimal expansions of an option.

Definition 3.2. Let $\mathcal{K}$ be a knowledge base and $\mathcal{O}$ an option for $\mathcal{K}$. We define the set of *minimal expansions* of $\mathcal{O}$ as follows:

$$exp(\mathcal{O}) = \{\mathcal{O}' \mid \mathcal{O}' \text{ is an option for } \mathcal{K} \wedge$$
$$\mathcal{O} \subset \mathcal{O}' \wedge$$
$$\text{there does not exist an option } \mathcal{O}'' \text{ for } \mathcal{K} \text{ s.t. } \mathcal{O} \subset \mathcal{O}'' \subset \mathcal{O}'\}$$

Given a set S of options, we define $exp(S) = \bigcup_{\mathcal{O} \in S} exp(\mathcal{O})$.

Clearly, the way $exp(\mathcal{O})$ is computed depends on the adopted weakening mechanism. In the following algorithm, the preference relation $\succeq$ is assumed to be anti-monotonic.

procedure CPO-Anti$(\mathcal{K}, \succeq)$
1. $S_0 = \{\mathcal{O} \mid \mathcal{O} \text{ is a minimal (under } \subseteq \text{) option for } \mathcal{K}\}$
2. Construct a maximal sequence $S_1, \ldots, S_n$ s.t. $S_i \neq \emptyset$ where
   $S_i = \{\mathcal{O} \mid \mathcal{O} \in exp(S_{i-1}) \wedge \nexists \mathcal{O}' \in S_0(\mathcal{O}' \subset \mathcal{O} \wedge \mathcal{O} \not\succeq \mathcal{O}')\}, 1 \leq i \leq n$
3. $S = \bigcup_{i=0}^{n} S_i$
4. Return the $\succeq$-preferred options in S

Clearly, the algorithm always terminates, since each option in S_i is a proper superset of some option in S_{i-1} and the size of an option for $\mathcal{K}$ is bounded. The algorithm exploits the anti-monotonicity of $\succeq$ to reduce the set of options from which the preferred ones are determined. First, the algorithm computes the minimal options for $\mathcal{K}$. Then, the algorithm computes bigger and bigger options and the anti-monotonicity of $\succeq$ is used to discard those options that are not preferred for sure: when S_i is computed, we consider every minimal expansion $\mathcal{O}$ of some option in S_{i-1}; if $\mathcal{O}$ is a proper superset of an option $\mathcal{O}' \in S_0$ and $\mathcal{O} \not\succeq \mathcal{O}'$, then $\mathcal{O}$ can be discarded since $\mathcal{O}' \succeq \mathcal{O}$ by the anti-monotonicity of $\succeq$ and therefore $\mathcal{O}' \succ \mathcal{O}$ (note that any option that is a superset of $\mathcal{O}$ will be discarded as well).

Observe that in the worst case the algorithm has to compute every option for $\mathcal{K}$ (e.g., when $\mathcal{O}_1 \succeq \mathcal{O}_2$ for any $\mathcal{O}_1, \mathcal{O}_2 \in \text{Opt}(\mathcal{K})$ as in this case every option is preferred).

Example 3.3. Consider the following knowledge base $\mathcal{K}$ containing check-in times for the employees in a company for a certain day.

ψ_1 *checkedIn_Mark_9AM*
ψ_2 *checkedIn_Claude_8AM*
ψ_3 *checkedIn_Mark_10AM*
ψ_4 $\neg(checkedIn\_Mark\_9AM \wedge checkedIn\_Mark\_10AM)$

Formula ψ_1 and ψ_2 state that employee *Mark* and *Claude* checked in for work at 9 and 8 AM, respectively. However, formula ψ_3 records that employee *Mark* checked in for work at 10 AM that day. Furthermore, as it is not possible for a person to check in for work at different times on the same day, we also have formula ψ_4, which is the instantiation of that constraint for employee *Mark*.

Assume that each formula ψ_i has an associated non-negative weight $w_i \in [0,1]$ corresponding to the likelihood of the formula being wrong, and suppose those weights are $w_1 = 0.2, w_2 = 0, w_3 = 0.1$, and $w_4 = 0$. Suppose that the weight of an option $\mathcal{O}$ is $w(\mathcal{O}) = \sum_{\psi_i \in \mathcal{K} \cap \mathcal{O}} w_i$. Let $\mathcal{W}_{\subseteq}$ be the weakening mechanism used, and consider the preference relation defined as follows: $\mathcal{O}_i \succeq \mathcal{O}_j$ iff $w(\mathcal{O}_i) \leq w(\mathcal{O}_j)$. Clearly, the preference relation is anti-monotonic. Algorithm **CPO-Anti** first computes $S_0 = \{\mathcal{O}_0 = \text{CN}(\emptyset)\}$. It then looks for the minimal expansions of $\mathcal{O}_0$ which are preferable to $\mathcal{O}_0$. In this case, we have $\mathcal{O}_1 = \text{CN}(\{\psi_2\})$ and $\mathcal{O}_2 = \text{CN}(\{\psi_4\})$; hence, $S_1 = \{\mathcal{O}_1, \mathcal{O}_2\}$. Note that neither $\text{CN}(\{\psi_1\})$ nor $\text{CN}(\{\psi_3\})$ is preferable to $\mathcal{O}_0$ and thus they can be discarded because $\mathcal{O}_0$ turns out to be strictly preferable to them. The algorithm then looks for the minimal expansions of some option in S_1 which are preferable to $\mathcal{O}_0$; the only one is $\mathcal{O}_3 = \text{CN}(\{\psi_2, \psi_4\})$, so $S_3 = \{\mathcal{O}_3\}$. It is easy to see that S_4 is empty and thus the algorithm returns the preferred options from those in $S_0 \cup S_1 \cup S_2 \cup S_3$, which are $\mathcal{O}_0$, $\mathcal{O}_1$, $\mathcal{O}_2$, and $\mathcal{O}_3$. Note that the algorithm avoided the computation of every option for $\mathcal{K}$.

We now show the correctness of the algorithm.

Theorem 3.3. *Let $\mathcal{K}$ be a knowledge base and $\succeq$ an anti-monotonic preference relation. Then,*

- *(Soundness) If **CPO-Anti**$(\mathcal{K}, \succeq)$ returns $\mathcal{O}$, then $\mathcal{O}$ is a preferred option for $\mathcal{K}$.*
- *(Completeness) For any preferred option $\mathcal{O}$ for $\mathcal{K}$, $\mathcal{O}$ is returned by* **CPO-Anti**$(\mathcal{K}, \succeq)$.

Proof. Let S be the set of options for $\mathcal{K}$ computed by the algorithm. First of all, we show that for any option $\mathcal{O}' \in \text{Opt}(\mathcal{K}) - S$ there exists an option $\mathcal{O}'' \in S$ s.t. $\mathcal{O}'' \succ \mathcal{O}'$. Suppose by contradiction that there is an option $\mathcal{O}' \in \text{Opt}(\mathcal{K}) - S$ s.t. there does not exist an option $\mathcal{O}'' \in S$ s.t. $\mathcal{O}'' \succ \mathcal{O}'$. Since $\mathcal{O}' \notin S_0$, then $\mathcal{O}'$ is not a minimal option for $\mathcal{K}$. Hence, there exist an option $\mathcal{O}_0 \in S_0$ and $n \geq 0$ options $\mathcal{O}_1, \ldots, \mathcal{O}_n$ s.t. $\mathcal{O}_0 \subset \mathcal{O}_1 \subset \ldots \subset \mathcal{O}_n \subset \mathcal{O}_{n+1} = \mathcal{O}'$ and $\mathcal{O}_i \in exp(\mathcal{O}_{i-1})$ for $1 \leq i \leq n+1$. Since $\nexists \mathcal{O}'' \in S_0$ s.t. $\mathcal{O}'' \succ \mathcal{O}'$, then $\nexists \mathcal{O}'' \in S_0$ s.t. $\mathcal{O}'' \succ \mathcal{O}_i$ for $0 \leq i \leq n$, otherwise $\mathcal{O}'' \succ \mathcal{O}_i$ and $\mathcal{O}_i \succeq \mathcal{O}'$ (by anti-monotonicity of $\succeq$) would imply $\mathcal{O}'' \succ \mathcal{O}'$, which is a contradiction. It can be easily verified, by induction on i, that $\mathcal{O}_i \in S_i$ for $0 \leq i \leq n+1$, and then $\mathcal{O}' \in S_{n+1}$, which is a contradiction.

(Soundness). Clearly, $\mathcal{O}$ is an option for $\mathcal{K}$. Suppose by contradiction that $\mathcal{O}$ is not preferred, i.e., there exists an option $\mathcal{O}'$ for $\mathcal{K}$ s.t. $\mathcal{O}' \succ \mathcal{O}$. Clearly, $\mathcal{O}' \in \text{Opt}(\mathcal{K}) - S$, otherwise it would be the case that $\mathcal{O}' \in S$ and then $\mathcal{O}$ is not returned by the algorithm (see step 4). We have proved above that there exists $\mathcal{O}'' \in S$ s.t. $\mathcal{O}'' \succ \mathcal{O}'$. Since $\mathcal{O}'' \succ \mathcal{O}'$ and $\mathcal{O}' \succ \mathcal{O}$, then $\mathcal{O}'' \succ \mathcal{O}$ (by the transitivity of $\succ$), which is a contradiction (as $\mathcal{O}, \mathcal{O}'' \in S$ and $\mathcal{O}$ is a $\succeq$-preferred option in S).

(Completeness). Suppose by contradiction that $\mathcal{O}$ is not returned by the algorithm. Clearly, this means that $\mathcal{O} \in \text{Opt}(\mathcal{K}) - S$. We have proved above that this implies that there exists an option $\mathcal{O}'' \in S$ s.t. $\mathcal{O}'' \succ \mathcal{O}$, which is a contradiction. $\square$

Observe that when the adopted weakening mechanism is either $\mathcal{W}_{\subseteq}$ or $\mathcal{W}_{all}$, the first step becomes $S_0 = \{CN(\emptyset)\}$, whereas the second step can be specialized as follows: $S_i = \{\mathcal{O} \mid \mathcal{O} \in exp(S_{i-1}) \wedge \mathcal{O} \succeq CN(\emptyset)\}$.

We now consider the case where $\succeq$ is assumed to be monotonic.

Definition 3.3. Let $\mathcal{K}$ be a knowledge base and $\mathcal{O}$ an option for $\mathcal{K}$. We define the set of *minimal contractions* of $\mathcal{O}$ as follows:

$$contr(\mathcal{O}) = \{\mathcal{O}' \mid \mathcal{O}' \text{ is an option for } \mathcal{K} \wedge$$
$$\mathcal{O}' \subset \mathcal{O} \wedge$$
$$\text{there does not exist an option } \mathcal{O}'' \text{ for } \mathcal{K} \text{ s.t. } \mathcal{O}' \subset \mathcal{O}'' \subset \mathcal{O}\}.$$

Given a set S of options, we define $contr(S) = \bigcup_{\mathcal{O} \in S} contr(\mathcal{O})$.

Observe that how to compute $contr(\mathcal{O})$ depends on the considered weakening mechanism. In the following algorithm the preference relation $\succeq$ is assumed to be monotonic.

procedure CPO-Monotonic$(\mathcal{K}, \succeq)$
1. $S_0 = \{\mathcal{O} \mid \mathcal{O} \text{ is a maximal (under } \subseteq) \text{ option for } \mathcal{K}\}$;
2. Construct a maximal sequence $S_1, \ldots, S_n$ s.t. $S_i \neq \emptyset$ where
   $S_i = \{\mathcal{O} \mid \mathcal{O} \in contr(S_{i-1}) \wedge \nexists \mathcal{O}' \in S_0(\mathcal{O} \subset \mathcal{O}' \wedge \mathcal{O} \not\succeq \mathcal{O}')\}, 1 \leq i \leq n$
3. $S = \bigcup_{i=0}^{n} S_i$
4. Return the $\succeq$-preferred options in S.

Clearly, the algorithm always terminates, since each option in S_i is a proper subset of some option in S_{i-1}. The algorithm exploits the monotonicity of $\succeq$ to reduce the set of options from which the preferred ones are determined. The algorithm first computes the maximal (under $\subseteq$) options for $\mathscr{K}$. It then computes smaller and smaller options and the monotonicity of $\succeq$ is used to discard those options that are not preferred for sure: when S_i is computed, we consider every minimal contraction $\mathcal{O}$ of some option in S_{i-1}; if $\mathcal{O}$ is a proper subset of an option $\mathcal{O}' \in S_0$ and $\mathcal{O} \not\succeq \mathcal{O}'$, then $\mathcal{O}$ can be discarded since $\mathcal{O}' \succeq \mathcal{O}$ by the monotonicity of $\succeq$ and therefore $\mathcal{O}' \succ \mathcal{O}$. Note that any option that is a subset of $\mathcal{O}$ will be discarded as well.

Observe that in the worst case the algorithm has to compute every option for $\mathscr{K}$ (e.g., when $\mathcal{O}_1 \succeq \mathcal{O}_2$ for any $\mathcal{O}_1, \mathcal{O}_2 \in \mathrm{Opt}(\mathscr{K})$ as in this case every option is preferred).

It is worth noting that when the adopted weakening mechanism is $\mathscr{W}_{all}$, the first step of the algorithm can be implemented by applying Theorem 3.1 since it identifies the options which are maximal under set inclusion (recall that $\succeq_W$ coincides with $\supseteq$, see Proposition 2.2). Likewise, when the weakening mechanism is $\mathscr{W}_{\subseteq}$, the first step of the algorithm can be accomplished by applying Corollary 3.1.

Example 3.4. Consider again the knowledge base of Example 3.3. Suppose now that each formula ψ_i has associated a non-negative weight $w_i \in [0,1]$ corresponding to the reliability of the formula, and let those weights be $w_1 = 0.1, w_2 = 1, w_3 = 0.2$, and $w_4 = 1$. Once again, the weight of an option $\mathcal{O}$ is $w(\mathcal{O}) = \sum_{\psi_i \in \mathscr{K} \cap \mathcal{O}} w_i$. Let $\mathscr{W}_{\subseteq}$ be the weakening mechanism, and consider the preference relation defined as follows: $\mathcal{O}_i \succeq \mathcal{O}_j$ iff $w(\mathcal{O}_i) \geq w(\mathcal{O}_j)$. Clearly, the preference relation is monotonic. Algorithm **CPO-Monotonic** first computes the maximal options, i.e., $S_0 = \{\mathcal{O}_1 = \mathrm{CN}(\{\psi_2, \psi_3, \psi_4\}), \mathcal{O}_2 = \mathrm{CN}(\{\psi_1, \psi_2, \psi_4\}), \mathcal{O}_4 = \mathrm{CN}(\{\psi_1, \psi_2, \psi_3\})\}$. After that, the algorithm looks for a minimal contraction $\mathcal{O}$ of some option in S_0 s.t. there is no superset $\mathcal{O}' \in S_0$ of $\mathcal{O}$ s.t. $\mathcal{O} \not\succeq \mathcal{O}'$. It is easy to see that in this case there is no option that satisfies this property, i.e., $S_1 = \emptyset$. Thus, the algorithm returns the preferred options in S_0, namely $\mathcal{O}_1$. Note that the algorithm avoided the computation of every option for $\mathscr{K}$.

We now show the correctness of the algorithm.

Theorem 3.4. *Let $\mathscr{K}$ be a knowledge base and $\succeq$ a monotonic preference relation. Then,*

- *(Soundness) If **CPO-Monotonic**$(\mathscr{K}, \succeq)$ returns $\mathcal{O}$, then $\mathcal{O}$ is a preferred option for $\mathscr{K}$.*
- *(Completeness) For any preferred option $\mathcal{O}$ for $\mathscr{K}$, $\mathcal{O}$ is returned by **CPO-Monotonic**$(\mathscr{K}, \succeq)$.*

Proof. Let S be the set of options for $\mathscr{K}$ computed by the algorithm. First of all, we show that for any option $\mathcal{O}' \in \mathrm{Opt}(\mathscr{K}) - S$, there exists an option $\mathcal{O}'' \in S$ s.t. $\mathcal{O}'' \succ \mathcal{O}'$. Suppose by contradiction that there is an option $\mathcal{O}' \in \mathrm{Opt}(\mathscr{K}) - S$ s.t. there does not exist an option $\mathcal{O}'' \in S$ s.t. $\mathcal{O}'' \succ \mathcal{O}'$. Since $\mathcal{O}' \not\subseteq S_0$, then $\mathcal{O}'$ is not a maximal option for $\mathscr{K}$. Hence, there exist an option $\mathcal{O}_0 \in S_0$ and $n \geq 0$ options

$\mathscr{O}_1, \ldots, \mathscr{O}_n$ s.t. $\mathscr{O}_0 \supset \mathscr{O}_1 \supset \ldots \supset \mathscr{O}_n \supset \mathscr{O}_{n+1} = \mathscr{O}'$ and $\mathscr{O}_i \in contr(\mathscr{O}_{i-1})$ for $1 \leq i \leq n+1$. Since $\not\exists \mathscr{O}'' \in S_0$ s.t. $\mathscr{O}'' \succ \mathscr{O}'$, then $\not\exists \mathscr{O}'' \in S_0$ s.t. $\mathscr{O}'' \succ \mathscr{O}_i$ for $0 \leq i \leq n$, otherwise $\mathscr{O}'' \succ \mathscr{O}_i$ and $\mathscr{O}_i \succeq \mathscr{O}'$ (by monotonicity of $\succeq$) would imply $\mathscr{O}'' \succ \mathscr{O}'$, which is a contradiction. It can be easily verified, by induction on i, that $\mathscr{O}_i \in S_i$ for $0 \leq i \leq n+1$, and then $\mathscr{O}' \in S_{n+1}$, which is a contradiction.

The soundness and completeness of the algorithm can be shown in the same way as in the proof of Theorem 3.3. $\square$

Chapter 4
Handling Inconsistency in Monotonic Logics

In this chapter, we analyze the usability of our framework for managing inconsistency in several monotonic logics, including ones for which inconsistency management has not been studied before (e.g., temporal, spatial, and probabilistic logics). We also study for some of these logics the computational complexity of obtaining the set of universal consequences for some specific combinations of weakening mechanisms and preference relations.

4.1 Propositional Horn-Clause Logic

Let us consider knowledge bases consisting of propositional Horn clauses. Recall that a Horn Clause is an expression of the form $L_1 \vee \cdots \vee L_n$ where each L_i is a propositional literal such that *at most* one L_i is positive.[1] We will assume that the consequences of a knowledge base are those determined by the application of modus ponens.

Proposition 4.1. *Consider a propositional Horn knowledge base $\mathscr{K}$. Let $\mathscr{W}_\subseteq$ and $\supseteq$ respectively be the weakening mechanism and preference relation that are used. A preferred option for $\mathscr{K}$ can be computed in polynomial time.*

Proof. Corollary 3.1 entails that a preferred option can be computed by finding a maximal consistent subset $\mathscr{K}'$ of $\mathscr{K}$ and then computing $CN(\mathscr{K}')$. Since both checking consistency and computing consequences can be accomplished in polynomial time (Papadimitriou 1994), the overall computation is in polynomial time. □

Nevertheless, the number of preferred options may be exponential, as shown in the following example.

---

[1] Note that a definite clause is a Horn clause where exactly one L_i is positive. It is well known that any set of definite clauses is always consistent.

Example 4.1. Consider the propositional Horn knowledge base

$$\mathcal{K} = \{a_1, \neg a_1, \ldots, a_n, \neg a_n\}$$

containing $2n$ formulas, where the a_i's are propositional variables. It is easy to see that the set of preferred options for $\mathcal{K}$ is

$$\{CN(\{l_1, \ldots, l_n\}) \mid l_i \in \{a_i, \neg a_i\} \text{ for } i = 1..n\}$$

whose cardinality is 2^n ($\mathcal{W}_\subseteq$ and $\supseteq$ are, respectively, the weakening mechanism and preference relation used).

The following theorem addresses the complexity of computing universal consequences of propositional Horn knowledge bases.

Proposition 4.2. *Let $\mathcal{K}$ and ψ be a propositional Horn knowledge base and clause, respectively. Let $\mathcal{W}_\subseteq$ and $\supseteq$ respectively be a weakening mechanism and preference relation. The problem of deciding whether ψ is a universal consequence of $\mathcal{K}$ is coNP-complete.*

Proof. It follows directly from Corollary 3.1 and the result that can be found in Cayrol and Lagasquie-Schiex (1994) stating that the problem of deciding whether a propositional Horn formula is a consequence of every maximal consistent subset of a Horn knowledge base is coNP-complete.   □

Note that when the weakening mechanism and the preference relation are $\mathcal{W}_{all}$ and $\succeq_W$, respectively, both the set of options and preferred options do not differ from those obtained when $\mathcal{W}_\subseteq$ and $\supseteq$ are considered. In fact, since $weakening(\psi) = \{\psi\}$ for any propositional Horn formula ψ, then $\mathcal{W}_\subseteq$ and $\mathcal{W}_{all}$ are the same. Proposition 2.2 states that $\supseteq$ and $\succeq_W$ coincide. Thus, the previous results are trivially extended to the case where $\mathcal{W}_{all}$ and $\succeq_W$ are considered.

Corollary 4.1. *Consider a propositional Horn knowledge base $\mathcal{K}$. Let $\mathcal{W}_{all}$ and $\succeq_W$ respectively be the weakening mechanism and preference relation that are used. A preferred option for $\mathcal{K}$ can be computed in polynomial time.*

Proof. Follows immediately from Proposition 4.1.   □

Corollary 4.2. *Let $\mathcal{K}$ and ψ be a propositional Horn knowledge base and clause, respectively. Let $\mathcal{W}_{all}$ and $\succeq_W$ respectively be the weakening mechanism and preference relation that are used. The problem of deciding whether ψ is a universal consequence of $\mathcal{K}$ is coNP-complete.*

Proof. Follows immediately from Proposition 4.2.   □

4.2 Propositional Probabilistic Logic

We now consider the probabilistic logic of Nilsson (1986) extended to probability intervals, i.e., formulas are of the form $\phi : [\ell, u]$, where ϕ is a classical propositional formula and $[\ell, u]$ is a subset of the real unit interval.

The existence of a set of propositional symbols is assumed. A *world* is any set of propositional symbols; we use W to denote the set of all possible worlds. A *probabilistic interpretation* I is a probability distribution over worlds, i.e., it is a function $I : W \rightarrow [0, 1]$ such that $\sum_{w \in W} I(w) = 1$. Then, I *satisfies* a formula $\phi : [\ell, u]$ iff $\ell \leq \sum_{w \in W, w \models \phi} I(w) \leq u$. Consistency and entailment are defined in the standard way.

Example 4.2. Consider a network of sensors collecting information about people's positions. Suppose the following knowledge base $\mathcal{K}$ is obtained by merging information collected by different sensors.

$$\psi_1 \quad loc\_John\_X : [0.6, 0.7]$$
$$\psi_2 \quad loc\_John\_X \lor loc\_John\_Y : [0.3, 0.5]$$

The first formula in $\mathcal{K}$ says that *John*'s position is X with a probability between 0.6 and 0.7. The second formula states that *John* is located either in position X or in position Y with a probability between 0.3 and 0.5. The knowledge base above is inconsistent: since every world in which the first formula is true satisfies the second formula as well, the probability of the latter has to be greater than or equal to the probability of the former.

As already mentioned before, a reasonable weakening mechanism for probabilistic knowledge bases consists of making probability intervals wider.

Definition 4.1. For any probabilistic knowledge base $\mathcal{K} = \{\phi_1 : [\ell_1, u_1], \ldots, \phi_n : [\ell_n, u_n]\}$, the weakening mechanism $\mathcal{W}_P$ is defined as follows: $\mathcal{W}_P(\mathcal{K}) = \{\{\phi_1 : [\ell_1', u_1'], \ldots, \phi_n : [\ell_n', u_n']\} \mid [\ell_i, u_i] \subseteq [\ell_i', u_i'], 1 \leq i \leq n\}$.

Example 4.3. Consider again the probabilistic knowledge base $\mathcal{K}$ of Example 4.2. The weakenings of $\mathcal{K}$ determined by $\mathcal{W}_P$ are of the form:

$$\psi_1' \quad loc\_John\_X : [\ell_1, u_1]$$
$$\psi_2' \quad loc\_John\_X \lor loc\_John\_Y : [\ell_2, u_2]$$

where $[0.6, 0.7] \subseteq [\ell_1, u_1]$ and $[0.3, 0.5] \subseteq [\ell_2, u_2]$. The options for $\mathcal{K}$ (w.r.t. $\mathcal{W}_P$) are the closure of those weakenings s.t. $[\ell_1, u_1] \cap [\ell_2, u_2] \neq \emptyset$ (this condition ensures consistency).

Suppose that the preferred options are those that modify the probability intervals as little as possible: $\mathcal{O}_i \succeq_P \mathcal{O}_j$ iff $sc(\mathcal{O}_i) \leq sc(\mathcal{O}_j)$ for any options $\mathcal{O}_i, \mathcal{O}_j$ for $\mathcal{K}$, where $sc(\mathrm{CN}(\{\psi_1', \psi_2'\})) = diff(\psi_1, \psi_1') + diff(\psi_2, \psi_2')$ and $diff(\phi : [\ell_1, u_1], \phi : [\ell_2, u_2]) = \ell_1 - \ell_2 + u_2 - u_1$. The preferred options are the closure of:

$$loc\_John\_X : [\ell, 0.7]$$
$$loc\_John\_X \lor loc\_John\_Y : [0.3, \ell]$$

where $0.5 \leq \ell \leq 0.6$.

We now define the preference relation introduced in the example above.

Definition 4.2. Let $\mathcal{K} = \{\phi_1 : [\ell_1, u_1], \ldots, \phi_n : [\ell_n, u_n]\}$ be a probabilistic knowledge base. We say that the score of an option $\mathcal{O} = CN(\{\phi_1 : [\ell'_1, u'_1], \ldots, \phi_n : [\ell'_n, u'_n]\})$ in $\mathsf{Opt}(\mathcal{K}, \mathcal{W}_P)$ is $sc(\mathcal{O}) = \sum_{i=1}^{n}(\ell_i - \ell'_i) + (u'_i - u_i)$. We define the preference relation $\succeq_P$ as follows: for any $\mathcal{O}, \mathcal{O}' \in \mathsf{Opt}(\mathcal{K}, \mathcal{W}_P)$, $\mathcal{O} \succeq_P \mathcal{O}'$ iff $sc(\mathcal{O}) \leq sc(\mathcal{O}')$.

The weakenings (under $\mathcal{W}_P$) whose closure yields the preferred options (w.r.t. $\succeq_P$) can be found by solving a linear program derived from the original knowledge base. We now show how to derive such a linear program.

In the following definition we use W to denote the set of possible worlds for a knowledge base $\mathcal{K}$, that is, $W = 2^{\Sigma}$, Σ being the set of propositional symbols appearing in $\mathcal{K}$.

Definition 4.3. Let $\mathcal{K} = \{\phi_1 : [\ell_1, u_1], \ldots, \phi_n : [\ell_n, u_n]\}$ be a probabilistic knowledge base. Then, $LP(\mathcal{K})$ is the following linear program:

$$\textbf{minimize } \sum_{i=1}^{n}(\ell_i - \ell'_i) + (u'_i - u_i)$$
$$\textbf{subject to}$$
$$\ell'_i \leq \sum_{w \in W, w \models \phi_i} p_w \leq u'_i, \quad 1 \leq i \leq n$$
$$\sum_{w \in W} p_w = 1$$
$$0 \leq \ell'_i \leq \ell_i, \quad 1 \leq i \leq n$$
$$u_i \leq u'_i \leq 1, \quad 1 \leq i \leq n$$

Clearly, in the definition above, the ℓ'_i's, u_i's and p_w's are variables (p_w denotes the probability of world w). We denote by $Sol(LP(\mathcal{K}))$ the set of solutions of $LP(\mathcal{K})$. We also associate a knowledge base $\mathcal{K}_{\mathscr{S}}$ to every solution $\mathscr{S}$ as follows: $\mathcal{K}_{\mathscr{S}} = \{\phi_i : [\mathscr{S}(\ell'_i), \mathscr{S}(u'_i)] \mid 1 \leq i \leq n\}$, where $\mathscr{S}(x)$ is the value assigned to variable x by solution $\mathscr{S}$. Intuitively, the knowledge base $\mathcal{K}_{\mathscr{S}}$ is the knowledge base obtained by setting the bounds of each formula in $\mathcal{K}$ to the values assigned by solution $\mathscr{S}$.

The following theorem states that the solutions of the linear program $LP(\mathcal{K})$ derived from a knowledge base $\mathcal{K}$ "correspond to" the preferred options of $\mathcal{K}$ when the weakening mechanism is $\mathcal{W}_P$ and the preference relation is $\succeq_P$.

Theorem 4.1. *Given a probabilistic knowledge base* $\mathcal{K}$,

1. If $\mathscr{S} \in Sol(LP(\mathcal{K}))$, *then* $\exists \mathcal{O} \in \mathsf{Opt}^{\succeq_P}(\mathcal{K}, \mathcal{W}_P)$ *s.t.* $\mathcal{O} = CN(\mathcal{K}_{\mathscr{S}})$,
2. If $\mathcal{O} \in \mathsf{Opt}^{\succeq_P}(\mathcal{K}, \mathcal{W}_P)$, *then* $\exists \mathscr{S} \in Sol(LP(\mathcal{K}))$ *s.t.* $\mathcal{O} = CN(\mathcal{K}_{\mathscr{S}})$.

Proof. Let LP' be the linear program obtained from $LP(\mathcal{K})$ by discarding the objective function.

(a) We first show that if $\mathscr{S} \in Sol(\text{LP}')$, then $\exists \mathscr{O} \in \text{Opt}(\mathscr{K}, \mathscr{W}_P)$ s.t. $\mathscr{O} = \text{CN}(\mathscr{K}_{\mathscr{S}})$. Clearly, $\mathscr{K}_{\mathscr{S}} \in \mathscr{W}_P(\mathscr{K})$ as the third and fourth sets of constraints in LP$'$ ensure that $[\ell_i, u_i] \subseteq [\ell_i', u_i']$ for any $\phi_i : [\ell_i, u_i] \in \mathscr{K}$. The first and second sets of constraints in LP$'$ ensure that $\mathscr{K}_{\mathscr{S}}$ is consistent – a model for $\mathscr{K}_{\mathscr{S}}$ is simply given by the p_w's. Thus, $\text{CN}(\mathscr{K}_{\mathscr{S}})$ is an option for $\mathscr{K}$.

(b) We now show that if $\mathscr{O} \in \text{Opt}(\mathscr{K}, \mathscr{W}_P)$, then $\exists \mathscr{S} \in Sol(\text{LP}')$ s.t. $\mathscr{O} = \text{CN}(\mathscr{K}_{\mathscr{S}})$. Since $\mathscr{O}$ is an option, then there exists $\mathscr{K}' \in \mathscr{W}_P(\mathscr{K})$ s.t. $\mathscr{O} = \text{CN}(\mathscr{K}')$. Clearly, $\mathscr{K}'$ is consistent. Let I be a model of $\mathscr{K}'$. It is easy to see that if we assign p_w to $I(w)$ for every world w and the ℓ_i''s and u_i''s are assigned to the bounds of ϕ_i in $\mathscr{K}'$, then such an assignment satisfies every constraint of LP$'$.

It is easy to see that given a solution $\mathscr{S}$ of LP$'$, the value of the objective function of $\text{LP}(\mathscr{K})$ for $\mathscr{S}$ is exactly the score sc assigned to the option $\text{CN}(\mathscr{K}_{\mathscr{S}})$ by $\succeq_P$ (see Definition 4.2).

1. Suppose that $\mathscr{S} \in Sol(\text{LP}(\mathscr{K}))$. As shown above, since $\mathscr{S}$ satisfies the constraints of $\text{LP}(\mathscr{K})$, then there exists an option $\mathscr{O}$ s.t. $\mathscr{O} = \text{CN}(\mathscr{K}_{\mathscr{S}})$. Suppose by contradiction that $\mathscr{O}$ is not preferred, that is, there is another option $\mathscr{O}'$ s.t. $sc(\mathscr{O}') < sc(\mathscr{O})$. Then, there is a solution $\mathscr{S}'$ of LP$'$ s.t. $\mathscr{O}' = \text{CN}(\mathscr{K}_{\mathscr{S}'})$. Since the objective function of $\text{LP}(\mathscr{K})$ corresponds to sc, then $\mathscr{S}$ does not minimize the objective function, which is a contradiction.

2. Suppose that $\mathscr{O} \in \text{Opt}^{\succeq_P}(\mathscr{K}, \mathscr{W}_P)$. As shown above, since $\mathscr{O}$ is an option, then there exists a solution $\mathscr{S}$ of LP$'$ s.t. $\mathscr{O} = \text{CN}(\mathscr{K}_{\mathscr{S}})$. Suppose by contradiction that $\mathscr{S}$ is not a solution of $\text{LP}(\mathscr{K})$. This means that it does not minimize the objective function. Then, there is a solution $\mathscr{S}'$ of $\text{LP}(\mathscr{K})$ which has a lower value of the objective function. As shown before, $\mathscr{O}' = \text{CN}(\mathscr{K}_{\mathscr{S}'})$ is an option and has a score lower than $\mathscr{O}$, which is a contradiction. $\qquad\square$

We refer to probabilistic knowledge bases whose formulas are built from propositional Horn formulas as *Horn probabilistic knowledge bases*. The following theorem states that already for this restricted subset of probabilistic logic, the problem of deciding whether a formula is a universal consequence of a knowledge base is coNP-hard.

Theorem 4.2. *Let $\mathscr{K}$ and ψ be a Horn probabilistic knowledge base and formula, respectively. Suppose that the weakening mechanism returns subsets of the given knowledge base and the preference relation is $\supseteq$. The problem of deciding whether ψ is a universal consequence of $\mathscr{K}$ is coNP-hard.*

Proof. We reduce 3-DNF VALIDITY to our problem. Let $\phi = C_1 \vee \ldots \vee C_n$ be an instance of 3-DNF VALIDITY, where the C_i's are conjunctions containing exactly three literals, and X the set of propositional variables appearing in ϕ. We derive from ϕ a Horn probabilistic knowledge base $\mathscr{K}^*$ as follows. Given a literal ℓ of the form x (resp. $\neg x$), with $x \in X$, we denote with $p(\ell)$ the propositional variable x^T (resp. x^F). Let

$$\mathscr{K}_1 = \{u \leftarrow p(\ell_1) \wedge p(\ell_2) \wedge p(\ell_3) : [1,1] \mid \ell_1 \wedge \ell_2 \wedge \ell_3 \text{ is a conjunction of } \phi\}$$

and

$$\mathcal{K}_2 = \{u \leftarrow x^T \wedge x^F : [1,1] \mid x \in X\}$$

Given a variable $x \in X$, let

$$\mathcal{K}_x = \{\; x^T : [1,1],$$
$$x^F : [1,1],$$
$$\leftarrow x^T \wedge x^F : [1,1]\}$$

Finally,

$$\mathcal{K}^* = \mathcal{K}_1 \cup \mathcal{K}_2 \cup \bigcup_{x \in X} \mathcal{K}_x$$

The derived instance of our problem is $(\mathcal{K}^*, u : [1,1])$. First of all, note that $\mathcal{K}^*$ is inconsistent since $\mathcal{K}_x$ is inconsistent for any $x \in X$. The set of maximal consistent subsets of $\mathcal{K}^*$ is:

$$\mathcal{M} = \left\{ \mathcal{K}_1 \cup \mathcal{K}_2 \cup \bigcup_{x \in X} \mathcal{K}_x' \mid \mathcal{K}_x' \text{ is a maximal consistent subset of } \mathcal{K}_x \right\}$$

Note that a maximal consistent subset of $\mathcal{K}_x$ is obtained from $\mathcal{K}_x$ by discarding exactly one formula. Corollary 3.1 entails that the set of preferred options for $\mathcal{K}^*$ is $\mathtt{Opt}^{\succeq}(\mathcal{K}^*) = \{\mathtt{CN}(S) \mid S \in \mathcal{M}\}$. We partition $\mathtt{Opt}^{\succeq}(\mathcal{K}^*)$ into two sets: $\mathcal{O}_1 = \{\mathcal{O} \mid \mathcal{O} \in \mathtt{Opt}^{\succeq}(\mathcal{K}^*) \wedge \exists x \in X \text{ s.t. } x^T : [1,1], x^F : [1,1] \in \mathcal{O}\}$ and $\mathcal{O}_2 = \mathtt{Opt}^{\succeq}(\mathcal{K}^*) - \mathcal{O}_1$. We now show that ϕ is valid iff $u : [1,1]$ is a universal consequence of $\mathcal{K}^*$.

($\Rightarrow$) It is easy to see that every preferred option $\mathcal{O}$ in $\mathcal{O}_1$ contains $u : [1,1]$, since there exists $x \in X$ s.t. $x^T : [1,1], x^F : [1,1] \in \mathcal{O}$ and $u \leftarrow x^T \wedge x^F : [1,1] \in \mathcal{O}$. Consider now a preferred option $\mathcal{O} \in \mathcal{O}_2$. For any $x \in X$ either $x^T : [1,1]$ or $x^F : [1,1]$ belongs to $\mathcal{O}$. Let us consider the truth assignment I derived from $\mathcal{O}$ as follows: for any $x \in X$, $I(x)$ is true iff $x^T : [1,1] \in \mathcal{O}$ and $I(x)$ is false iff $x^F : [1,1] \in \mathcal{O}$. Since ϕ is valid, then I satisfies ϕ, i.e., there is a conjunction $\ell_1 \wedge \ell_2 \wedge \ell_3$ of ϕ which is satisfied by I. It is easy to see that $u : [1,1]$ can be derived from the rule $u \leftarrow p(\ell_1) \wedge p(\ell_2) \wedge p(\ell_3) : [1,1]$ in $\mathcal{K}_1$. Hence, $u : [1,1]$ is a universal consequence of $\mathcal{K}^*$.

($\Leftarrow$) We show that if ϕ is not valid then there exists a preferred option $\mathcal{O}$ for $\mathcal{K}^*$ s.t. $u : [1,1] \notin \mathcal{O}$. Consider a truth assignment for ϕ which does not satisfy ϕ and let $True$ and $False$ be the set of variables of ϕ made true and false, respectively, by such an assignment. Consider the following set

$$S = \mathcal{K}_1 \cup \mathcal{K}_2$$
$$\cup \bigcup_{x \in True} \{x^T : [1,1], \leftarrow x^T \wedge x^F : [1,1]\}$$
$$\cup \bigcup_{x \in False} \{x^F : [1,1], \leftarrow x^T \wedge x^F : [1,1]\}$$

It is easy to see that S is a maximal consistent subset of $\mathcal{K}^*$, and thus $\mathcal{O} = \mathtt{CN}(S)$ is a preferred option for $\mathcal{K}^*$. It can be easily verified that $u : [1,1] \notin \mathcal{O}$. $\square$

4.3 Propositional Linear Temporal Logic

Temporal logic has been extensively used for reasoning about programs and their executions. It has achieved a significant role in the formal specification and verification of concurrent and distributed systems (Pnueli 1977). In particular, a number of useful concepts such as safety, liveness and fairness can be formally and concisely specified using temporal logics (Manna and Pnueli 1992; Emerson 1990).

In this section, we consider *Propositional Linear Temporal Logic* (PLTL) (Gabbay et al. 1980) – a logic used in verification of systems and reactive systems. Basically, this logic extends classical propositional logic with a set of temporal connectives. The particular variety of temporal logic we consider is based on a linear, discrete model of time isomorphic to the natural numbers. Thus, the temporal connectives operate over a sequence of distinct "moments" in time. The connectives that we consider are $\Diamond$ (*sometime in the future*), $\Box$ (*always in the future*) and $\bigcirc$ (*at the next point in time*).

Assuming a countable set Σ of propositional symbols, every $p \in \Sigma$ is a PLTL formula. If ϕ and ψ are PLTL formulas, then the following are PLTL formulas as well: $\phi \vee \psi$, $\phi \wedge \psi$, $\neg\phi$, $\phi \leftarrow \psi$, $\Box\phi$, $\bigcirc\phi$, $\Diamond\phi$

The notion of a timeline can be formalized with a function $I : \mathbb{N} \to 2^{\Sigma}$ that maps each natural number (representing a moment in time) to a set of propositional symbols (intuitively, this is the set of propositional symbols which are true at that moment). We say that

- $(I,i) \models p$ iff $p \in I(i)$, where $p \in \Sigma$;
- $(I,i) \models \bigcirc\phi$ iff $(I,i+1) \models \phi$;
- $(I,i) \models \Diamond\phi$ iff $\exists j. j \geq i \wedge (I,j) \models \phi$;
- $(I,i) \models \Box\phi$ iff $\forall j. j \geq i \, implies \, (I,j) \models \phi$.

The semantics for the standard connectives is as expected. I is a model of a PLTL formula ϕ iff $(I,0) \models \phi$. Consistency and entailment are defined in the standard way.

Example 4.4. Consider the PLTL knowledge base reported below (Artale 2008) which specifies the behavior of a computational system.

$$\psi_1 \quad \Box(\Diamond received \leftarrow requested)$$
$$\psi_2 \quad \Box(\bigcirc processed \leftarrow received)$$

The first statement says that it is always the case that if a request is issued, then it will be received at some future time point. The second statement says that it is always the case that if a request is received, then it will be processed at the next time point. The statements above correspond to the definition of the system, i.e., how the system is supposed to behave. Suppose now that there is a monitoring system which reports data regarding the system's behavior and, for instance, the following formula is added to the knowledge base:

$$\psi_3 \quad received \wedge \bigcirc\neg processed \wedge \bigcirc\bigcirc processed$$

The inclusion of ψ_3 makes the knowledge base inconsistent, since the monitoring system is reporting that a request was received and was not processed at the next moment in time, but two moments afterwards.

Consider a weakening function that replaces the $\bigcirc$ operator with the $\Diamond$ operator in a formula ψ, provided that the formula thus obtained is a consequence of ψ. Suppose that preferred options are those that keep as many monitoring system formulas unchanged (i.e. unweakened) as possible. In this case, the only preferred option is $CN(\{\psi_1, \Box(\Diamond processed \leftarrow received), \psi_3\})$, where formula ψ_2, stating that if a request is received then it will be processed at the next moment in time, has been weakened into a new one stating that the request will be processed at a future point in time.

Theorem 4.3. *Let $\mathcal{K}$ and ψ be a temporal knowledge base and formula, respectively. Suppose the weakening mechanism returns subsets of the given knowledge base and the preference relation is $\supseteq$. The problem of deciding whether ψ is a universal consequence of $\mathcal{K}$ is coNP-hard.*

Proof. A reduction from 3-DNF VALIDITY to our problem can be carried out in a similar way to the proof of Theorem 4.2. Let $\phi = C_1 \vee \ldots \vee C_n$ be an instance of 3-DNF VALIDITY, where the C_i's are conjunctions containing exactly three literals, and X the set of propositional variables appearing in ϕ. We derive from ϕ a temporal knowledge base $\mathcal{K}^*$ as follows. Given a literal ℓ of the form x (resp. $\neg x$), with $x \in X$, we denote with $p(\ell)$ the propositional variable x^T (resp. x^F). Let

$$\mathcal{K}_1 = \{\Box(u \leftarrow p(\ell_1) \wedge p(\ell_2) \wedge p(\ell_3)) \mid \ell_1 \wedge \ell_2 \wedge \ell_3 \text{ is a conjunction of } \phi\}$$

and

$$\mathcal{K}_2 = \{\Box(u \leftarrow x^T \wedge x^F) \mid x \in X\}$$

Given a variable $x \in X$, let

$$\mathcal{K}_x = \{\ \Box x^T, \\ \Box x^F, \\ \Box(\leftarrow x^T \wedge x^F)\}$$

Finally,

$$\mathcal{K}^* = \mathcal{K}_1 \cup \mathcal{K}_2 \cup \bigcup_{x \in X} \mathcal{K}_x$$

The derived instance of our problem is $(\mathcal{K}^*, \Box u)$. The claim can be proved in a similar way to the proof of Theorem 4.2. $\qquad\qquad\square$

4.4 Fuzzy Logic

In this section we consider fuzzy logic. Formulas are of the form $\phi : v$, where ϕ is a propositional formula built from a set Σ of propositional symbols and the logical connectives $\neg, \wedge, \vee$, and $v \in [0, 1]$ (we call v the *degree of truth*). An interpretation

I assigns a value in $[0,1]$ to each propositional symbol in Σ. Moreover, given two propositional formulas ϕ_1 and ϕ_2, we have

- $I(\neg\phi_1) = 1 - I(\phi_1)$
- $I(\phi_1 \wedge \phi_2) = min\{I(\phi_1), I(\phi_2)\}$
- $I(\phi_1 \vee \phi_2) = max\{I(\phi_1), I(\phi_2)\}$

We say that I satisfies a formula $\phi : v$ iff $I(\phi) \geq v$. Consistency and entailment are defined in the standard way.

Example 4.5. Consider the following inconsistent knowledge base:

$$\psi_1 : a : 0.7$$
$$\psi_2 : b : 0.6$$
$$\psi_3 : \neg(a \wedge b) : 0.5$$

Suppose that the weakening mechanism is defined as follows: for any formula $\phi : v$, we define $\mathscr{W}(\phi : v) = \{\phi : v' \mid v' \in [0,1] \wedge v' \leq v\}$; then, $\mathscr{W}(\mathscr{K}) = \{\{\psi_1', \psi_2', \psi_3'\} \mid \psi_i' \in \mathscr{W}(\psi_i)\ 1 \leq i \leq 3\}$. Thus, options are the closure of

$$\psi_1 : a : v_1$$
$$\psi_2 : b : v_2$$
$$\psi_3 : \neg(a \wedge b) : v_3$$

where $v_1 \leq 0.7$, $v_2 \leq 0.6$, $v_3 \leq 0.5$, and $1 - min\{v_1, v_2\} \geq v_3$. Suppose that the preferred options are those that modify a minimum number of formulas. Thus, the preferred options are of the form

$$CN(\{a : v_1, \psi_2, \psi_3\})\ \text{where}\ v_1 \leq 0.5, or$$
$$CN(\{b : v_2, \psi_1, \psi_3\})\ \text{where}\ v_2 \leq 0.5, or$$
$$CN(\{\neg(a \wedge b) : v_3, \psi_1, \psi_2\})\ \text{where}\ v_3 \leq 0.4$$

Finally, suppose that the preference relation is expressed as before but, in addition, we would like to change the degree of truth as little as possible. In this case, the preferred options are:

$$CN(\{b : 0.5, \psi_1, \psi_3\})$$
$$CN(\{\neg(a \wedge b) : 0.4, \psi_1, \psi_2\})$$

We refer to fuzzy knowledge bases whose formulas are built from propositional Horn formulas as *Horn fuzzy knowledge bases*. The following theorem states that even for this restricted subset of fuzzy logic, the problem of deciding whether a formula is a universal consequence of a knowledge base is coNP-hard.

Theorem 4.4. *Let $\mathscr{K}$ and ψ be a Horn fuzzy knowledge base and formula, respectively. Let $\mathscr{W}_\subseteq$ and $\supseteq$ be the adopted weakening mechanism and preference relation, respectively. The problem of deciding whether ψ is a universal consequence of $\mathscr{K}$ is coNP-hard.*

Proof. A reduction from 3-DNF VALIDITY to our problem can be carried out in a way similar to the proof of Theorem 4.2. Let $\phi = C_1 \vee \ldots \vee C_n$ be an instance of 3-DNF VALIDITY, where the C_i's are conjunctions containing exactly three literals, and X is the set of propositional variables appearing in ϕ. We derive from ϕ a Horn temporal knowledge base $\mathscr{K}^*$ as follows. Given a literal ℓ of the form x (resp. $\neg x$), with $x \in X$, we denote with $p(\ell)$ the propositional variable x^T (resp. x^F). Let

$$\mathscr{K}_1 = \{u \vee \neg p(\ell_1) \vee \neg p(\ell_2) \vee \neg p(\ell_3) : 1 \mid \ell_1 \wedge \ell_2 \wedge \ell_3 \text{ is a conjunction of } \phi\}$$

and

$$\mathscr{K}_2 = \{u \vee \neg x^T \vee \neg x^F : 1 \mid x \in X\}$$

Given a variable $x \in X$, let

$$\mathscr{K}_x = \{\ x^T : 1,$$
$$x^F : 1,$$
$$\neg x^T \vee \neg x^F : 1\}$$

Finally,

$$\mathscr{K}^* = \mathscr{K}_1 \cup \mathscr{K}_2 \cup \bigcup_{x \in X} \mathscr{K}_x$$

The derived instance of our problem is $(\mathscr{K}^*, u : 1)$. The claim can be proved in a way similar to the proof of Theorem 4.2. $\square$

4.5 Belief Logic

In this section we focus on the belief logic presented in Levesque (1984). Formulas are formed from a set Σ of primitive propositions, the standard connectives $\vee$, $\wedge$, and $\neg$, and two unary connectives B and L. Neither B nor L appear within the scope of the other. Connective B is used to express what is *explicitly* believed by an agent (a sentence that is actively held to be true by the agent), whereas L is used to express what is *implicitly* believed by the agent (i.e., the consequences of his explicit beliefs).

   Semantics of sentences is given in terms of a model structure $\langle S, \mathscr{B}, T, F \rangle$, where S is a set of *situations*, $\mathscr{B}$ is a subset of S (the situations that could be the actual ones according to what is believed), and T and F are functions from Σ to subsets of S. Intuitively, $T(p)$ are the situations that support the truth of p and $F(p)$ are the situations that support the falsity of p. A primitive proposition may be true, false, both, or neither in a situation. A *complete* situation (or *possible world*) is one that supports either the truth or the falsity (not both) of every primitive proposition. A complete situation s is *compatible* with a situation s' if s and s' agree whenever s' is defined, i.e., if $s' \in T(p)$ then $s \in T(p)$, and if $s' \in F(p)$ then $s \in F(p)$, for each primitive proposition p. Let $\mathscr{W}(\mathscr{B})$ consist of all complete situations in S compatible with some situation in $\mathscr{B}$.

Two *support relations* $\models_T$ and $\models_F$ between situations and formulas are defined in the following way:

- $s \models_T p$ iff $s \in T(p)$, where p is a primitive proposition ,
- $s \models_F p$ iff $s \in F(p)$, where p is a primitive proposition ;
- $s \models_T (\alpha \vee \beta)$ iff $s \models_T \alpha$ or $s \models_T \beta$,
- $s \models_F (\alpha \vee \beta)$ iff $s \models_F \alpha$ and $s \models_F \beta$;
- $s \models_T (\alpha \wedge \beta)$ iff $s \models_T \alpha$ and $s \models_T \beta$,
- $s \models_F (\alpha \wedge \beta)$ iff $s \models_F \alpha$ or $s \models_F \beta$;
- $s \models_T \neg\alpha$ iff $s \models_F \alpha$,
- $s \models_F \neg\alpha$ iff $s \models_T \alpha$;
- $s \models_T B\alpha$ iff for every $s' \in \mathscr{B}, s' \models_T \alpha$,
- $s \models_F B\alpha$ iff $s \not\models_T B\alpha$;
- $s \models_T L\alpha$ iff for every $s' \in \mathscr{W}(\mathscr{B})$, $s' \models_T \alpha$,
- $s \models_F L\alpha$ iff $s \not\models_T L\alpha$.

Given a complete situation s in S, then if $s \models_T \alpha$, then α is true at s, otherwise α is said to be false at s. Finally, α is said to be valid ($\models \alpha$) iff for any model structure $\langle S, B, T, F \rangle$ and any complete situation s in S, α is true at s. The satisfiability of a sentence is defined analogously; entailment is defined in the expected way.

Note that belief logic allows an agent to believe contradictory sentences, e.g., $\{Bp, B\neg p\}$ is a consistent knowledge base. However, $\{Bp, \neg Bp\}$ is inconsistent.

Example 4.6. Consider the following inconsistent knowledge base $\mathscr{K}$ that represents the knowledge of an agent regarding a city's subway system:

$\psi_1 : goingNorthTrain1$
$\psi_2 : B\,goingNorthTrain1$
$\psi_3 : goingNorthTrain1 \rightarrow canGetUpTownFromStationA$
$\psi_4 : B(goingNorthTrain1 \rightarrow canGetUpTownFromStationA)$
$\psi_5 : \neg(canGetUpTownFromStationA)$

Using a train schedule associated with train station A, we might be able to express formulas ψ_1 and ψ_3. ψ_1 states that *Train 1* goes north, whereas ψ_3 states that if *Train 1* goes north, then the agent can get uptown from station A. Formulas ψ_2 and ψ_4 state that the agent explicitly believes in the information that he got from the schedule. However, this knowledge base is inconsistent because of the presence of formula ψ_5, which states that it is not possible to get uptown from station A, for instance, because that route is closed for repairs.

Suppose that each formula ψ_i is associated with a time stamp $t(\psi_i)$ that represents the moment in time in which the agent acquired that piece of information. In this case, we consider the subsets of $\mathscr{K}$ as its weakenings, and the preference relation is defined in such a way that maximal (under $\subseteq$) options are preferable to the others, and among these we say that $\mathscr{O}_i \succeq \mathscr{O}_j$ iff $sc(\mathscr{O}_i) \geq sc(\mathscr{O}_j)$ where $sc(\mathscr{O}) = \sum_{\psi \in \mathscr{O} \cap \mathscr{K}} t(\psi)$, i.e., we would like to preserve as many formulas as possible and more up to date information. If in our example we have $t(\psi_1) = t(\psi_2) = 1, t(\psi_3) = t(\psi_4) = 3$, and $t(\psi_5) = 5$, then the only preferred option is $\mathsf{CN}(\{\psi_2, \psi_3, \psi_4, \psi_5\})$.

Theorem 4.5. *Let $\mathcal{K}$ and ψ be a belief knowledge base and formula, respectively. Suppose that the weakening mechanism returns subsets of the given knowledge base and the preference relation is $\supseteq$. The problem of deciding whether ψ is a universal consequence of $\mathcal{K}$ is coNP-hard.*

Proof. A reduction from 3-DNF VALIDITY to our problem can be carried out in a similar way to the proof of Theorem 4.2 by using only propositional formulas. Let $\phi = C_1 \vee \ldots \vee C_n$ be an instance of 3-DNF VALIDITY, where the C_i's are conjunctions containing exactly three literals, and X is the set of propositional variables appearing in ϕ. We derive from ϕ a belief knowledge base $\mathcal{K}^*$ as follows. Given a literal ℓ of the form x (resp. $\neg x$), with $x \in X$, we denote by $p(\ell)$ the propositional variable x^T (resp. x^F). Let

$$\mathcal{K}_1 = \{ u \vee \neg p(\ell_1) \vee \neg p(\ell_2) \vee \neg p(\ell_3) \mid \ell_1 \wedge \ell_2 \wedge \ell_3 \text{ is a conjunction of } \phi \}$$

and

$$\mathcal{K}_2 = \{ u \vee \neg x^T \vee \neg x^F \mid x \in X \}$$

Given a variable $x \in X$, let

$$\mathcal{K}_x = \{ \; x^T, \\ x^F, \\ \neg x^T \vee \neg x^F \}$$

Finally,

$$\mathcal{K}^* = \mathcal{K}_1 \cup \mathcal{K}_2 \cup \bigcup_{x \in X} \mathcal{K}_x$$

The derived instance of our problem is $(\mathcal{K}^*, u)$. The claim can be proved in a similar way to the proof of Theorem 4.2. $\qquad\square$

4.6 Spatial Logic

In this section, we consider the Region Connection Calculus (RCC) proposed in Randell et al. (1992). This logic is a topological approach to qualitative spatial representation and reasoning (see Cohn and Renz 2008) where *spatial regions* are subsets of a topological space. Relationships between spatial regions are defined in first order logic in terms of a primitive binary relation $C(x, y)$, which means "x connects with y"; this relation is reflexive and symmetric and holds when regions x and y share a common point. **RCC-8** considers the following eight *base relations*: *DC* (*disconnected*), *EC* (*externally connected*), *PO* (*partial overlap*), *EQ* (*equal*), *TPP* (*tangential proper part*), *NTPP* (*non-tangential proper part*), *TPP*$^{-1}$ (*the inverse of TPP*), *NTPP*$^{-1}$ (*the inverse of NTPP*). These relations are jointly exhaustive and pairwise disjoint, i.e., exactly one of them holds between any two spatial regions. Figure 4.1 shows two-dimensional examples of the eight basic relations. If the exact

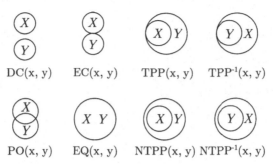

Fig. 4.1 Two-dimensional examples for **RCC-8** base relations

relation is not known, we can also use the union of different base relations, e.g., $a\{NTPP,NTPP^{-1}\}b$ means that either a is a non-tangential proper part of b or vice versa.

Example 4.7. Consider the following knowledge base $\mathscr{K}$:

$$
\begin{array}{ll}
\psi_1 & a\{NTPP\}b \\
\psi_2 & b\{EC\}c \\
\psi_3 & a\{EC\}c
\end{array}
$$

The knowledge base is inconsistent since the first two formulas imply that a and c are disconnected, whereas the last one states that they are externally connected.

Suppose the weakening mechanism used is $\mathscr{W}_{all}$. In this case, the knowledge base is weakened by making its formulas more undefined. For instance, some options for $\mathscr{K}$ are

$$
\begin{array}{l}
\mathscr{O}_1 = \mathrm{CN}(\{a\{NTPP,TPP\}b, \psi_2, \psi_3\}) \\
\mathscr{O}_2 = \mathrm{CN}(\{b\{EC,PO,NTPP^{-1}\}c, \psi_1, \psi_3\}) \\
\mathscr{O}_3 = \mathrm{CN}(\{b\{EC,NTPP^{-1}\}c, \psi_1, \psi_3\}) \\
\mathscr{O}_4 = \mathrm{CN}(\{a\{NTPP,TPP\}b, b\{EC,DC\}c, \psi_3\})
\end{array}
$$

Suppose the preference relation chooses those options that weaken a minimum number of formulas as the preferred options. In this case, $\mathscr{O}_1$, $\mathscr{O}_2$, $\mathscr{O}_3$ are preferred options, whereas $\mathscr{O}_4$ is not.

Suppose the preference relation is $\succeq_W$. Then $\mathscr{O}_1$ and $\mathscr{O}_3$ are preferred options, whereas $\mathscr{O}_2$ and $\mathscr{O}_4$ are not. In fact, it is easy to see that $\mathscr{O}_3 \succeq_W \mathscr{O}_2$ but not vice versa, and $\mathscr{O}_1 \succeq_W \mathscr{O}_4$ but not vice versa.

Finally, suppose that options that only weaken formulas of the form $x\{NTPP\}y$ or $x\{TPP\}y$ into $x\{NTPP,TPP\}y$ are preferable to the others (e.g., because we are not sure if a region is a tangential or non-tangential proper part of another, but the information about the other topological relations is reliable and we would prefer not to change it). In this case, $\mathscr{O}_1$ is a preferred option whereas the others are not.

Chapter 5
Link with Existing Approaches

Many important works in inconsistency management have been developed in AI literature in the last three decades. In this chapter we revisit some of the most influential works in this area and analyze how our general framework relates to them, showing in some cases how the proposals correspond to special cases of our framework by defining adequate weakening mechanisms and preference relations. Two kinds of approaches have been proposed in the literature for solving the problem of inconsistency. The first type focuses on revising the knowledge base and restoring its consistency. The second approach accepts inconsistency and copes with it, prohibiting the logic from deriving trivial inferences.

The first approach, initiated in Rescher and Manor (1970), proposes to give up some formulas of the knowledge base in order to get one or several consistent subbases. More specifically, Rescher and Manor (1970) considers propositional knowledge bases. Preferences among the maximal consistent subsets of the original knowledge may be expressed, so that *preferred* maximal consistent subsets are determined. A formula is a *P-consequence* of a possibly inconsistent knowledge base $\mathcal{K}$, where P is the preference criterion, if it is a consequence of every preferred (according to P) maximal consistent subset of $\mathcal{K}$. The paper discusses various preference criteria. Given a knowledge base $\mathcal{K}$ and a preference criterion P, we use $MCS(\mathcal{K}, P)$ to denote the set of preferred maximal consistent subsets of $\mathcal{K}$.

Definition 5.1. Consider a knowledge base $\mathcal{K}$ and a preference criterion P on the maximal consistent subsets of $\mathcal{K}$. Suppose $\mathcal{W}_{\subseteq}$ is the weakening mechanism used. For any $\mathcal{O}_1, \mathcal{O}_2 \in \mathsf{Opt}(\mathcal{K})$, we say that $\mathcal{O}_1 \succeq \mathcal{O}_2$ iff there exists $\mathcal{K}_1 \in MCS(\mathcal{K}, P)$ s.t. $\mathcal{O}_1 = \mathsf{CN}(\mathcal{K}_1)$.

As stated in the following proposition, the preferred maximal consistent subsets *correspond to* the preferred options when the weakening mechanism is $\mathcal{W}_{\subseteq}$ and the preference relation is the one of the definition above.

Proposition 5.1. *Let $\mathcal{K}$ be a knowledge base and P a preference criterion on the maximal consistent subsets of $\mathcal{K}$. Let $\mathcal{W}_{\subseteq}$ be the adopted weakening mechanism and $\succeq$ the preference relation of Definition 5.1. Then:*

M.V. Martinez et al., *A General Framework for Reasoning On Inconsistency*, SpringerBriefs in Computer Science, DOI 10.1007/978-1-4614-6750-2_5, © The Author(s) 2013

- $\forall S \in MCS(\mathcal{K},P),\ \exists \mathcal{O} \in \text{Opt}^{\succeq}(\mathcal{K})$ *such that* $\mathcal{O} = \text{CN}(S)$.
- $\forall \mathcal{O} \in \text{Opt}^{\succeq}(\mathcal{K}),\ \exists S \in MCS(\mathcal{K},P)$ *such that* $\mathcal{O} = \text{CN}(S)$.

Proof. Straightforward.   □

Clearly, *P*-consequences correspond to our notion of universal consequence (Definition 2.8). Note that our framework is not restricted to propositional knowledge bases only and gives the flexibility to choose a weakening mechanism different from $\mathcal{W}_{\subseteq}$.

In the case of prioritized knowledge bases, Brewka (1989) has proposed a definition of a *preferred subbase*. The basic idea is to take as much important information into account as possible. More specifically, two generalizations of Poole's approach (Poole 1988) have been proposed. In the following, a knowledge base is a set of classical first order formulas. In the first generalization, a knowledge base $\mathcal{K}$ is supposed to be stratified into $\mathcal{K}_1, \ldots, \mathcal{K}_n$ ($\mathcal{K} = \mathcal{K}_1 \cup \ldots \cup \mathcal{K}_n$) such that the formulas in the same stratum are equally preferred, whereas formulas in a stratum $\mathcal{K}_i$ are preferred to formulas in $\mathcal{K}_j$ with $i < j$.

Definition 5.2. (Brewka 1989) Let $\mathcal{K} = \mathcal{K}_1 \cup \ldots \cup \mathcal{K}_n$ be a knowledge base. $S = S_1 \cup \ldots \cup S_n$ is a *preferred subbase* of $\mathcal{K}$ if and only if $\forall j, 1 \leq j \leq n, S_1 \cup \ldots \cup S_j$ is a maximal (under set-inclusion) consistent subset of $\mathcal{K}_1 \cup \ldots \cup \mathcal{K}_j$.
$P_1(\mathcal{K})$ denotes the set of preferred subbases of $\mathcal{K}$.

We show that the approach above can be captured by our framework by defining the appropriate weakening mechanism and preference relation.

Definition 5.3. Consider a knowledge base $\mathcal{K}$ and let $\mathcal{W}_{\subseteq}$ be the adopted weakening mechanism. For any $\mathcal{O}_1, \mathcal{O}_2 \in \text{Opt}(\mathcal{K})$, we say that $\mathcal{O}_1 \succeq \mathcal{O}_2$ iff there exists $\mathcal{K}_1 \in P_1(\mathcal{K})$ s.t. $\mathcal{O}_1 = \text{CN}(\mathcal{K}_1)$.

Proposition 5.2. *Let $\mathcal{K}$ be a knowledge base, $\mathcal{W}_{\subseteq}$ the weakening mechanism and $\succeq$ the preference relation of Definition 5.3. Then,*

- $\forall S \in P_1(\mathcal{K}),\ \exists \mathcal{O} \in \text{Opt}^{\succeq}(\mathcal{K})$ *such that* $\mathcal{O} = \text{CN}(S)$.
- $\forall \mathcal{O} \in \text{Opt}^{\succeq}(\mathcal{K}),\ \exists S \in P_1(\mathcal{K})$ *such that* $\mathcal{O} = \text{CN}(S)$.

Proof. Straightforward.   □

The second generalization is based on a partial order on the formulas of a knowledge base.

Definition 5.4. Let $<$ be a strict partial order on a knowledge base $\mathcal{K}$. S is a *preferred subbase* of $\mathcal{K}$ if and only if there exists a strict total order $\psi_1, \ldots, \psi_n$ of $\mathcal{K}$ respecting $<$ such that $S = S_n$ with

$$S_0 = \emptyset$$
$$S_i = \begin{cases} S_{i-1} \cup \{\psi_i\} & \textit{if } S_{i-1} \cup \{\psi_i\} \textit{ is consistent} \\ S_{i-1} & \textit{otherwise} \end{cases} \quad 1 \leq i \leq n$$

$P_2(\mathcal{K})$ denotes the set of preferred subbases of $\mathcal{K}$.

In addition, the second generalization can be easily expressed in our framework.

Definition 5.5. Consider a knowledge base $\mathcal{K}$ and let $\mathcal{W}_{\subseteq}$ be the adopted weakening mechanism. For any $\mathcal{O}_1, \mathcal{O}_2 \in \mathrm{Opt}(\mathcal{K})$, we say that $\mathcal{O}_1 \succeq \mathcal{O}_2$ iff there exists $\mathcal{K}_1 \in P_2(\mathcal{K})$ s.t. $\mathcal{O}_1 = \mathrm{CN}(\mathcal{K}_1)$.

Proposition 5.3. *Let $\mathcal{K}$ be a knowledge base, $\mathcal{W}_{\subseteq}$ the weakening mechanism used, and $\succeq$ the preference relation of Definition 5.5. Then,*

- *$\forall S \in P_2(\mathcal{K})$, $\exists \mathcal{O} \in \mathrm{Opt}^{\succeq}(\mathcal{K})$ such that $\mathcal{O} = \mathrm{CN}(S)$.*
- *$\forall \mathcal{O} \in \mathrm{Opt}^{\succeq}(\mathcal{K})$, $\exists S \in P_2(\mathcal{K})$ such that $\mathcal{O} = \mathrm{CN}(S)$.*

Proof. Straightforward.   □

Brewka (1989) provides a *weak* and *strong* notion of provability for both the generalizations described above. A formula ψ is *weakly provable* from a knowledge base $\mathcal{K}$ iff there is a preferred subbase S of $\mathcal{K}$ s.t. $\psi \in \mathrm{CN}(S)$; ψ is *strongly provable* from $\mathcal{K}$ iff for every preferred subbase S of $\mathcal{K}$ we have $\psi \in \mathrm{CN}(S)$. Clearly, the latter notion of provability corresponds to our notion of universal consequence (Definition 2.8), whereas the former is not a valid inference mechanism, since the set of weakly provable formulas might be inconsistent. Observe that Brewka's approach is committed to a specific logic, weakening mechanism and preference criterion, whereas our framework is applicable to different logics and gives the flexibility to choose the weakening mechanism and the preference relation that the end-user believes more suitable for his purposes.

Inconsistency management based on a partial order on the formulas of a knowledge base has also been studied in Nico (1992). In this work the author defines the concept of a *reliability theory*, based on a partial *reliability relation* among the formulas in a first order logic knowledge base $\mathcal{K}$. In terms of this theory the set of all *most reliable consistent set of premises* is defined, which corresponds to $P_2(\mathcal{K})$ in Brewka's approach (Brewka 1989). The set of theorems that can be proved from the theory is the set of propositions that are logically entailed by every most reliable consistent set, which basically corresponds to our set of universal consequences. Clearly, this approach can be expressed in our framework in a manner analogous to Definition 5.5 for Brewka's approach. The author defines a special purpose logic based on first order calculus, and a deduction process to obtain the set of premises that can be *believed* to be true. The deduction process is based on the computation of justifications (premises used in the derivation of contradictions) for believing or removing formulas, and it iteratively constructs and refines these justifications. At each step, the set of formulas that can be believed from a set of justifications can be computed in time $O(n * m)$ where n is the number of justifications used in that step and m is the number of formulas in the theory.

Priority-based management of inconsistent knowledge bases has been addressed also in Benferhat et al. (1993) (see also Cayrol and Lagasquie-Schiex 1995). Specifically, propositional knowledge bases are considered and a knowledge base $\mathcal{K}$ is supposed to be stratified into strata $\mathcal{K}_1, \ldots, \mathcal{K}_n$, where $\mathcal{K}_1$ consists of the formulas of highest priority whereas $\mathcal{K}_n$ contains the formulas of lowest priority. Priorities in $\mathcal{K}$ are used to select preferred consistent subbases. Inferences are made from the preferred subbases of $\mathcal{K}$, that is $\mathcal{K}$ entails a formula ψ iff ψ can be classically inferred from every preferred subbase of $\mathcal{K}$. The paper presents different meaning of "preferred", which are reported in the following definition.

Definition 5.6. (Benferhat et al. 1993) Let $\mathcal{K} = (\mathcal{K}_1 \cup \ldots \cup \mathcal{K}_n)$ be a propositional knowledge base, $X = (X_1 \cup \ldots \cup X_n)$ and $Y = (Y_1 \cup \ldots \cup Y_n)$ two consistent subbases of $\mathcal{K}$, where $X_i = X \cap \mathcal{K}_i$ and $Y_i = Y \cap \mathcal{K}_i$. We define:

- *Best-out preference*: let $a(Z) = min\{i \mid \exists \psi \in \mathcal{K}_i - Z\}$ for a consistent subbase Z of $\mathcal{K}$, with the convention $min\emptyset = n + 1$. The best-out preference is defined by $X \ll^{bo} Y$ iff $a(X) \le a(Y)$;
- *Inclusion-based preference* is defined by $X \ll^{incl} Y$ iff $\exists i$ s.t. $X_i \subset Y_i$ and $\forall j$ s.t. $1 \le j < i, X_j = Y_j$;
- *Lexicographic preference* is defined by $X \ll^{lex} Y$ iff $\exists i$ s.t. $|X_i| < |Y_i|$ and $\forall j$ s.t. $1 \le j < i, |X_j| = |Y_j|$.

Let us consider the best-out preference and let $amax(\mathcal{K}) = max\{i \mid \mathcal{K}_1 \cup \ldots \cup \mathcal{K}_i \text{ is consistent}\}$. If $amax(\mathcal{K}) = k$, then the *best-out preferred* consistent subbases of $\mathcal{K}$ are exactly the consistent subbase of $\mathcal{K}$ which contain $(\mathcal{K}_1 \cup \ldots \cup \mathcal{K}_k)$; we denote them by $P_{bo}(\mathcal{K})$. This approach can be easily captured by our framework by adopting $\mathcal{W}_\subseteq$ as weakening mechanism and defining the preference relation as follows.

Definition 5.7. Consider a knowledge base $\mathcal{K}$ and let $\mathcal{W}_\subseteq$ be the adopted weakening mechanism. For any $\mathcal{O}_1, \mathcal{O}_2 \in \mathtt{Opt}(\mathcal{K})$, we say that $\mathcal{O}_1 \succeq \mathcal{O}_2$ iff there exists $\mathcal{K}_1 \in P_{bo}(\mathcal{K})$ s.t. $\mathcal{O}_1 = \mathtt{CN}(\mathcal{K}_1)$.

Proposition 5.4. *Let $\mathcal{K}$ be a knowledge base, $\mathcal{W}_\subseteq$ the weakening mechanism and $\succeq$ the preference relation of Definition 5.7. Then,*

- $\forall S \in P_{bo}(\mathcal{K})$, $\exists \mathcal{O} \in \mathtt{Opt}^{\succeq}(\mathcal{K})$ *such that* $\mathcal{O} = \mathtt{CN}(S)$.
- $\forall \mathcal{O} \in \mathtt{Opt}^{\succeq}(\mathcal{K})$, $\exists S \in P_{bo}(\mathcal{K})$ *such that* $\mathcal{O} = \mathtt{CN}(S)$.

Proof. Straightforward.   □

The *inclusion-based preferred* subbases are of the form $(X_1 \cup \ldots \cup X_n)$ s.t. $(X_1 \cup \ldots \cup X_i)$ is a maximal (under set inclusion) consistent subbase of $(\mathcal{K}_1 \cup \ldots \cup \mathcal{K}_i)$, for $i = 1..n$. Note these preferred subbases coincide with Brewka's preferred subbases of Definition 5.2 above, which can be expressed in our framework.

Finally, the lexicographic preferred subbases are of the form $(X_1 \cup \ldots \cup X_n)$ s.t. $(X_1 \cup \ldots \cup X_i)$ is a cardinality-maximal consistent subbase of $(\mathcal{K}_1 \cup \ldots \cup \mathcal{K}_i)$, for $i = 1..n$; we denote them by $P_{lex}(\mathcal{K})$.

Definition 5.8. Consider a knowledge base $\mathcal{K}$ and let $\mathcal{W}_\subseteq$ be the adopted weakening mechanism. For any $\mathcal{O}_1, \mathcal{O}_2 \in \mathtt{Opt}(\mathcal{K})$, we say that $\mathcal{O}_1 \succeq \mathcal{O}_2$ iff there exists $\mathcal{K}_1 \in P_{lex}(\mathcal{K})$ s.t. $\mathcal{O}_1 = \mathtt{CN}(\mathcal{K}_1)$.

Proposition 5.5. *Let $\mathcal{K}$ be a knowledge base, $\mathcal{W}_\subseteq$ the weakening mechanism and $\succeq$ the preference relation of Definition 5.8. Then,*

- $\forall S \in P_{lex}(\mathcal{K})$, $\exists \mathcal{O} \in \mathsf{Opt}^{\succeq}(\mathcal{K})$ *such that* $\mathcal{O} = \mathsf{CN}(S)$.
- $\forall \mathcal{O} \in \mathsf{Opt}^{\succeq}(\mathcal{K})$, $\exists S \in P_{lex}(\mathcal{K})$ *such that* $\mathcal{O} = \mathsf{CN}(S)$.

Proof. Straightforward. □

As already said before, once a criterion for determining preferred subbase has been fixed, a formula is a consequence of $\mathcal{K}$ if can be classically inferred from every preferred subbase, which corresponds to our universal inference mechanism (Definition 2.8).

In Cayrol and Lagasquie-Schiex (1995), the same criteria for selecting preferred consistent subbases are considered, and three entailment principles are presented. The *UNI principle* corresponds to our universal inference mechanism and it is the same as in Benferhat et al. (1993). According to the *EXI principle*, a formula ψ is inferred from a knowledge base $\mathcal{K}$ if ψ is classically inferred from at least one preferred subbase of $\mathcal{K}$. According to the *ARG principle*, a formula ψ is inferred from a knowledge base $\mathcal{K}$ if ψ is classically inferred from at least one preferred subbase and no preferred subbase classically entails $\neg\psi$. The last two entailment principles are not valid inference mechanisms in our framework, since the set of EXI (resp. ARG) consequences might be inconsistent.

The second approach to handling inconsistency does not aim to restore consistency, instead the main objective is to reason despite the inconsistent information and treats inconsistent information as informative. Argumentation and paraconsistent logic are examples of this approach (we refer the reader to Hunter (1998) for a survey on paraconsistent logics). Argumentation is based on the justification of plausible conclusions by arguments (Amgoud and Cayrol 2002; Dung 1995). Due to inconsistency, arguments may be attacked by counterarguments. The problem is thus to select the most acceptable arguments. Several semantics were proposed in the literature for that purpose (see Baroni et al. 2011 for a survey).

In Amgoud and Besnard (2010), an argumentation system was proposed for reasoning about inconsistent premises. The system is grounded on Tarski's logics. It builds arguments from a knowledge base as follows:

Definition 5.9 (Argument). Let $\mathcal{K}$ be a knowledge base. An *argument* is a pair (X,x) s.t. $X \subseteq \mathcal{K}$, X is consistent, $x \in \mathsf{CN}(X)$ and $\nexists X' \subset X$ such that $x \in \mathsf{CN}(X')$. X is called the *support* of the argument and x its *conclusion*.

Notations: For $\mathcal{K} \subseteq \mathcal{L}$, $\mathsf{Arg}(\mathcal{K})$ is the set of all arguments that may be built from $\mathcal{K}$ using Definition 5.9. For an argument $a = (X,x)$, $\mathsf{Conc}(a) = x$ and $\mathsf{Supp}(a) = X$.

An argument a attacks an argument b if it undermines one of its premises.

Definition 5.10 (Attack). Let a,b be two arguments. a attacks b, denoted $a\mathcal{R}_u b$, iff $\exists y \in \mathsf{Supp}(b)$ such that the set $\{\mathsf{Conc}(a), y\}$ is inconsistent.

An argumentation system over a given knowledge base is thus the pair: set of arguments and the attacks among them.

Definition 5.11 (Argumentation system). An *argumentation system* (AS) over a knowledge base $\mathcal{K}$ is a pair $\mathcal{T} = (\text{Arg}(\mathcal{K}), \mathcal{R}_u)$.

It is worth mentioning that the set $\text{Arg}(\mathcal{K})$ may be infinite even when the base $\mathcal{K}$ is finite. This would mean that the argumentation system may be *infinite*.[1]
Arguments are evaluated using stable semantics that was proposed in Dung (1995).

Definition 5.12 (Stable semantics). Let $\mathcal{T} = (\text{Arg}(\mathcal{K}), \mathcal{R}_u)$ be an AS over a knowledge base $\mathcal{K}$, and $\mathcal{E} \subseteq \text{Arg}(\mathcal{K})$ s.t. $\nexists\ a, b \in \mathcal{E}$ s.t. $a\mathcal{R}_u b$. The set $\mathcal{E}$ is a *stable* extension iff $\forall a \in \text{Arg}(\mathcal{K}) \setminus \mathcal{E}$, $\exists b \in \mathcal{E}$ s.t. $b\mathcal{R}_u a$.
Let $\text{Ext}(\mathcal{T})$ be the set of stable extensions of $\mathcal{T}$.

The conclusions that may be drawn from $\mathcal{K}$ by an argumentation system are the formulas that are conclusions of arguments in each stable extension.

Definition 5.13 (Output). Let $\mathcal{T} = (\text{Arg}(\mathcal{K}), \mathcal{R}_u)$ be an AS over a knowledge base $\mathcal{K}$. For $x \in \mathcal{L}$, $\mathcal{K} \hspace{1pt}\vert\hspace{-2pt}\sim x$ iff $\forall \mathcal{E} \in \text{Ext}(\mathcal{T})$, $\exists a \in \mathcal{E}$ s.t. $\text{Conc}(a) = x$. $\text{Output}(\mathcal{T}) = \{x \in \mathcal{L} \mid \mathcal{K} \hspace{1pt}\vert\hspace{-2pt}\sim x\}$.

In Amgoud (2012), it was shown that this argumentation system captures the approach of Rescher and Manor recalled earlier in this chapter. Indeed, each stable extension returns a maximal (under set inclusion) consistent subbase of the original knowledge base $\mathcal{K}$. Moreover, the plausible conclusions drawn from $\mathcal{K}$ are those inferred from all the maximal subbases of $\mathcal{K}$. In light of Proposition 5.1, it is easy to show that the abstract framework proposed in this book captures also the argumentation system proposed in Amgoud and Besnard (2010).

Proposition 5.6. *Let $\mathcal{K}$ be a knowledge base and P a preference criterion on the maximal consistent subsets of $\mathcal{K}$. Let $\mathcal{W}_\subseteq$ be the adopted weakening mechanism and $\succeq$ the preference relation of Definition 5.1. Let $\mathcal{T} = (\text{Arg}(\mathcal{K}), \mathcal{R}_u)$ be an AS over the base $\mathcal{K}$.*

- $\forall \mathcal{E} \in \text{Ext}(\mathcal{T})$, $\exists \mathcal{O} \in \text{Opt}^{\succeq}(\mathcal{K})$ *such that* $\mathcal{O} = \text{CN}(\bigcup_{a \in \mathcal{E}} \text{Supp}(a))$.
- $\forall \mathcal{O} \in \text{Opt}^{\succeq}(\mathcal{K})$, $\exists \mathcal{E} \in \text{Ext}(\mathcal{T})$ *such that* $\mathcal{O} = \text{CN}(\bigcup_{a \in \mathcal{E}} \text{Supp}(a))$.

Proof. Straightforward. $\square$

---

[1] An AS is *finite* iff each argument is attacked by a finite number of arguments. It is *infinite* otherwise.

Chapter 6
Conclusions

Past works on reasoning about inconsistency in AI have suffered from multiple flaws: (i) they apply to one logic at a time and are often invented for one logic after another. (ii) They assume that the AI researcher will legislate how applications resolve inconsistency even though the AI researcher may often know nothing about a specific application which may be built in a completely different time frame and geography than the AI researcher's work – in the real world, users are often stuck with the consequences of their decisions and would often like to decide what they want to do with their data (including what data to consider and what not to consider when there are inconsistencies). An AI system for reasoning about inconsistent information must support the user in his/her needs rather than forcing something down their throats. (iii) Most existing frameworks use some form or the other of maximal consistent subsets.

In this monograph, we attempt to address all these three flaws through a single unified approach that builds upon Tarksi's axiomatization of what a logic is. Most existing monotonic logics such as classical logic, Horn logic, probabilistic logic, temporal logic are special cases of Tarski's definition of a logic. Thus, we develop a framework for reasoning about inconsistency in any logic that satisfies Tarski's axioms. Second, we propose the notion of an "option" in any logic satisfying Tarski's axioms. An option is a set of formulas in the logic that is closed and consistent – however, the end user is not forced to choose a maximal consistent subset and options need not be maximal or even subsets of the original inconsistent knowledge base. Another element of our framework is that of preference. Users can specify any preference relation they want on their options.

Once the user has selected the logic he is working with, the options that he considers appropriate, and his preference relation on these options, our framework provides a semantics for a knowledge base *taking these user inputs into account*.

Our framework for reasoning about inconsistency has three basic components: (i) a set of options which are consistent and closed sets of formulas determined from the original knowledge base by means of a weakening mechanism which is general enough to apply to arbitrary logics and that allows users to flexibly specify how to weaken a knowledge base according to their application domains

M.V. Martinez et al., *A General Framework for Reasoning On Inconsistency*, SpringerBriefs in Computer Science, DOI 10.1007/978-1-4614-6750-2_6, © The Author(s) 2013

and needs. (ii) A general notion of preference relation between options. We show that our framework not only captures maximal consistent subsets, but also many other criteria that a user may use to select between options. We have also shown that by defining an appropriate preference relation over options, we can capture several existing works such as the subbases defined in Rescher and Manor (1970) and Brewka's subtheories. (iii) The last component of the framework consists of an inference mechanism that allows the selection of the inferences to be drawn from the knowledge base. This mechanism should return an option. This forces the system to make safe inferences.

We have also shown through examples how this abstract framework can be used in different logics, provided new results on the complexity of reasoning about inconsistency in such logics, and proposed general algorithms for computing preferred options.

In short, our framework empowers end-users to make decisions about what they mean by an option, what options they prefer to what other options, and prevents them from being dragged down by some systemic assumptions made by a researcher who might never have seen their application or does not understand the data and/or the risks posed to the user in decision making based on some a priori definition of what data should be discarded when an inconsistency arises.

References

Amgoud L (2012) Stable semantics in logic-based argumentation. In: International conference on scalable uncertainty management (SUM), Marburg, pp 58–71

Amgoud L, Besnard P (2010) A formal analysis of logic-based argumentation systems. In: International conference on scalable uncertainty management (SUM), Toulouse, pp 42–55

Amgoud L, Cayrol C (2002) Inferring from inconsistency in preference-based argumentation frameworks. J Autom Reason 29(2):125–169

Artale A (2008) Formal methods: linear temporal logic. http://www.inf.unibz.it/~artale/FM/slide3.pdf

Bacchus F (1990) Representing and reasoning with probabilistic knowledge. MIT, Cambridge

Baroni P, Caminada M, Giacomin M (2011) An introduction to argumentation semantics. Knowl Eng Rev 26(4):365–410

Benferhat S, Cayrol C, Dubois D, Lang J, Prade H (1993) Inconsistency management and prioritized syntax-based entailment. In: International joint conference on artificial intelligence (IJCAI), Chambéry, pp 640–647

Bobrow DG (1980) Special issue on non-monotonic reasoning. Artif Intell J 13(1–2)

Brewka G (1989) Preferred subtheories: an extended logical framework for default reasoning. In: International joint conference on artificial intelligence (IJCAI), Detroit, pp 1043–1048

Cayrol C, Lagasquie-Schiex M (1994) On the complexity of non-monotonicentailment in syntax-based approaches. In: ECAI workshop on algorithms, complexity and commonsense reasoning

Cayrol C, Lagasquie-Schiex M (1995) Non-monotonic syntax-based entailment: a classification of consequence relations. In: Symbolic and quantitative approaches to reasoning and uncertainty (ECSQARU), Fribourg, pp 107–114

Cohn AG, Renz J (2008) Qualitative spatial representation and reasoning. In: van Hermelen F, Lifschitz V, Porter B (eds) Handbook of knowledge representation. Elsevier, Amsterdam/Boston, pp 551–596

M.V. Martinez et al., *A General Framework for Reasoning On Inconsistency*, SpringerBriefs in Computer Science, DOI 10.1007/978-1-4614-6750-2, © The Author(s) 2013

Dung PM (1995) On the acceptability of arguments and its fundamental role in nonmonotonic reasoning, logic programming and n-person games. Artif Intell 77:321–357

Emerson EA (1990) Temporal and modal logic. In: Handbook of theoretical computer science. Elsevier, Amsterdam/New York, pp 995–1072

Gabbay DM, Pnueli A, Shelah S, Stavi J (1980) On the temporal basis of fairness. In: Symposium on principles of programming languages (POPL), Las Vegas, pp 163–173

Gardenfors P (1988) The dynamics of belief systems: foundations vs. coherence. Int J Philos

Halpern JY (1990) An analysis of first-order logics of probability. Artif Intell 46(3):311–350

Hunter A (1998) Paraconsistent logics. In: Besnard P, Gabbay DM, Hunter A, Smets P (eds) Handbook of defeasible reasoning and uncertainty management systems, volume 2: reasoning with actual and potential contradictions, Kluwer, Dordrecht, pp 11–36

Levesque HJ (1984) A logic of implicit and explicit belief. In: National conference on artificial intelligence (AAAI), Austin, pp 198–202

Manna Z, Pnueli A (1992) The temporal logic of reactive and concurrent systems: specification. Springer, New York

Nico R (1992) A logic for reasoning with inconsistent knowledge. Artif Intell 57(1):69–103

Nilsson NJ (1986) Probabilistic logic. Artif Intell 28(1):71–87

Papadimitriou CH (1994) Computational complexity. Addison-Wesley, Reading

Pinkas G, Loui RP (1992) Reasoning from inconsistency: a taxonomy of principles for resolving conflicts. In: International conference on principles of knowledge representation and reasoning (KR), Cambridge, pp 709–719

Pnueli A (1977) The temporal logic of programs. In: Symposium on foundations of computer science (FOCS), Providence, pp 46–57

Poole D (1985) On the comparison of theories: preferring the most specific explanation. In: International joint conference on artificial intelligence (IJCAI), Los Angeles, pp 144–147

Poole D (1988) A logical framework for default reasoning. Artif Intell 36(1):27–47

Randell DA, Cui Z, Cohn AG (1992) A spatial logic based on regions and connection. In: Principles of knowledge representation and reasoning (KR), Cambridge, pp 165–176

Reiter R (1980) A logic for default reasoning. Artif Intell 13:81–132

Rescher N, Manor R (1970) On inference from inconsistent premises. Theory Decis 1:179–219

Shoenfield J (1967) Mathematical logic. Addison-Wesley, Reading

Subrahmanian VS, Amgoud L (2007) A general framework for reasoning about inconsistency. In: International joint conference on artificial intelligence (IJCAI), Hyderabad, pp 599–504

Tarski A (1956) On some fundamental concepts of metamathematics, Oxford University Press, London

Touretzkey DS (1984) Implicit ordering of defaults in inheritance systems. In: National conference on artificial intelligence (AAAI), Austin, pp 322–325

Index

M.V. Martinez et al., *A General Framework for Reasoning On Inconsistency*, SpringerBriefs   45
in Computer Science, DOI 10.1007/978-1-4614-6750-2, © The Author(s) 2013